ADVANCE PRAISE FOR *THE GERM FILES*

"Jason Tetro has provided an owner's manual for the trillions of personal organisms that we cannot live without. It's an easy-read guide to how they participate in health, food, beauty, and even sex. It will change the way you look at your entire being."
—Bob McDonald, host of
CBC Radio's *Quirks & Quarks*

"Reading this book is like going on safari with a smart and experienced adventure guide. Sit back as Tetro brings into focus a world of magnificent microscopic creatures, and reveals what their fascination behaviour means for human well-being.
—Ziya Tong, host of *Daily Planet*, Discovery Channel

"When you flush the toilet, keep the lid down. If you want fresh breath, chew a stick of gum for two minutes. *The Germ Files* is filled with incredibly practical advice like that, backed by the latest scientific evidence. Jason Tetro curates a breezy take on the bacteria, viruses, fungi and other little beasties among us. But behind the easy read is an important message. Some germs are our best friends. Others are our worst enemies. Knowing which is which makes your life better and might one day even save it."

—Brian Goldman, host of
CBC Radio's *White Coat, Black Art*

PRAISE FOR *THE GERM CODE*

"A refreshing and unique perspective on a complicated, hidden world. Whereas most accounts of germs dwell on the grisly details of the infections they cause, Jason Tetro reminds us that the reality is far more nuanced, and that in many respects our lives and theirs are entwined in a delicately balanced dance. *The Germ Code* is a highly readable and enjoyable overview of that relationship."

—Dr. Peter Hotez, Dean, National School
of Tropical Medicine, Baylor College of Medicine

"With *The Germ Code*, first-time author Jason Tetro has delivered an important book that brings the world of microorganisms entertainingly to life. Tetro has a breezy style and relishes the interesting tale. . . . There are more than enough absorbing stories to keep the reader hooked."

—*Toronto Star*

THE GERM FILES

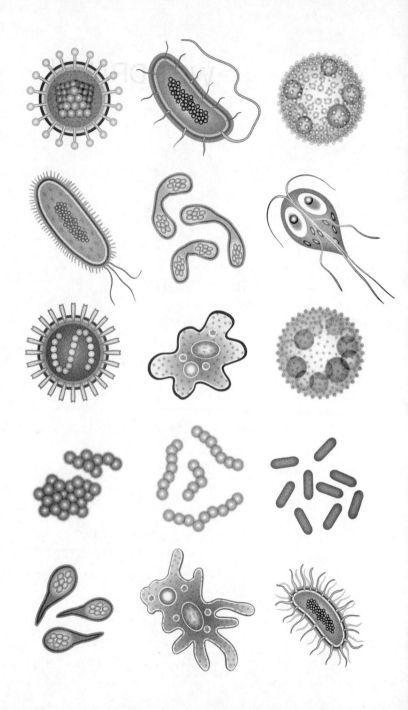

JASON TETRO

THE

THE SURPRISING WAYS MICROBES CAN

GERM

IMPROVE YOUR HEALTH AND LIFE

FILES

(AND HOW TO PROTECT YOURSELF FROM THE BAD ONES)

DOUBLEDAY CANADA

Doubleday Canada and colophon are registered
trademarks of Penguin Random House Canada Limited

Library and Archives Canada Cataloguing in Publication

Tetro, Jason, author
The germ files : the surprising ways microbes can improve your health and life
(and how to protect yourself from the bad ones) / Jason Tetro.

Issued in print and electronic formats.
ISBN 978-0-385-68577-1 (paperback).—ISBN 978-0-385-68578-8 (epub)
1. Bacteria—Popular works. 2. Bacteria—Social aspects—
Popular works. 3. Germ theory of disease—Popular works.
4. Microbiology—Popular works. I. Title.
QR56.T483 2016   579.3   C2015-906191-1
C2015-906192-X

Book design: CS Richardson
Cover and interior illustrations: Macrovector/Shutterstock.com
Printed and bound in the USA

Published in Canada by Doubleday Canada,
a division of Penguin Random House Canada Limited

www.penguinrandomhouse.ca

10 9 8 7 6 5 4 3 2

Penguin
Random House
DOUBLEDAY CANADA

To Anastassia Voronova

Contents

INTRODUCTION

ARE YOU A GERMOPHOBE? Do you fear the microbes that occupy every inch of our planet? Does the mere mention of bacteria, viruses, fungi, or amoebae make you want to find the nearest sink basin or hand-sanitizer station?

This fear is entirely natural given that germs have been considered our enemies for over a century. We've declared war on them, developed entire industries to combat them, and devoted ample time and space in the media to describe how they harm human health. Look at the American reaction to the Ebola scare of 2014. Although the actual risk of contracting the disease was limited to a small room in a Dallas hospital, an entire country was frozen in fear.

It may not seem so from this example and several others—from the flu pandemic of 2010 to the appearance of measles at Disneyland—but minds are changing. Germs are no longer viewed solely as a scourge. Thanks to the work of tens of thousands of researchers worldwide, we have gained a new appreciation for these creatures. These excellent scientists regularly venture into the microscopic world to observe the many varieties of germs and how they influence our health and our environment. Although there are some bad actors, most inhabitants of the microbial world mean us no harm, and may even benefit our lives.

Over most of the past three decades, I have been fortunate to be able to take part in this never-ending fact-finding mission. My fascination when I first started studying germs as a teenager was unbridled and has never waned. Whenever I think I have seen it all—every manner of incredible microbial phenomenon—a new trait of a bacterium, virus, fungi, protozoa, or worm is discovered, amazing me as a scientist and sustaining my personal curiosity and my hunger to know more. When this new-found characteristic somehow relates to the lives and welfare of human beings, my fascination is unconfined.

The source of this wonder comes down to a basic but incredibly important fact: we need germs to live. They surround us and outnumber us, and thousands of species call our bodies their home. There are far more good germs than bad—only 0.1 percent of the thousands of species regularly interacting with us are known to cause infection—but our focus nonetheless tends to be on the latter. After all, it is far more interesting to discuss a killer than a helper.

In the past decade, we've gained quite a bit of insight into how the friendly microbes affect all aspects of our lives and our health. We now know how hundreds of species interact with us, and more importantly, what their presence can mean for our health, our relationships, and . . . well, the way we smell. But the information is rather one-way and tends not to acknowledge that we too can influence our microbes to help keep us in the best of health.

My goal when I began writing this book was to ensure that every piece within it would have some bearing on our daily lives. Whereas *The Germ Code* was a narrative about our involuntary marriage with germs, *The Germ Files* is more of a guide to improving the relationship. You can jump in at any point and learn about the incredible role microbes play in our lives, as well as discovering ways to bring harmony between the body and the bugs. And the information is well-founded, on the conclusions of well over 1,000 scientific articles, chapters and reports.

As a final note—if you are a germophobe, I truly hope you read this book. I promise the content will not make the situation worse. Indeed, I hope the information will comfort you and convince you that for the most part, germs are not out to get us; they want to live with us. If we do our part, we can make the relationship work to our mutual benefit.

Volvox berberi

1. MEET YOUR GERMS

A CAST OF TRILLIONS Here's an easy experiment to better understand the impact of microbes on our lives. Head to a mirror and stand before it. How many living organisms do you see in the reflection? Most people will say, "One." Humans have a tendency to think of themselves as individuals.

Now, ask the same question of a cellular biologist and the reply will be quite different: about 37 trillion. For that person, every cell in the body is a living organism, capable of surviving and thriving on its own. The human body is a collection of these cells, all working together to form who we are.

Ask a microbiologist—like me—and the answer nearly quadruples, to 137 trillion. To us, human cells make up only a fraction of the living organisms co-habiting with the person in the mirror. Most of the residents—up to 100 trillion—are microbial organisms, including bacteria, viruses, fungi, amoebae, and worms.

The idea of all those germs moving about inside and on you might make you squirm, but it's a perfectly natural situation. Microbes are everywhere. Swab almost any place on earth and you will be sure to find some proof of microbial life. The same goes for our bodies. Bacteria can be found on our skin as well as in our mouths, sinuses, guts, bladders, breasts, blood and eyes. This is your microbial population.

For years, researchers have worked hard to identify all the microbes calling the human body home. The task is daunting, to say the least; the number of different microbial types, also called species, can be in the millions. But it seems that only about ten thousand reside with humans, with anywhere from five hundred to over a thousand found in or on any one person. Each species amasses in large numbers, usually in the hundreds of billions, to form collectives, or colonies. Some even establish areas of residence called biofilms. There, they can enjoy the good life and thrive.

But knowing the number is only half of the work. In biological sciences, there is another essential question to be asked about any living creature: Why is it there? Understanding the function of an organism, both as an individual and as part of a collective, is just as important as knowing it exists at all.

Let's put this into context. The phrase "37 trillion human cells" refers to a number—how many there are. But numbers alone don't tell us very much. In order to understand the importance of these cells, we need to know their function.

Human cells can be grouped into thirteen different types, including bone, nerve, muscle, blood, skin, and fat. The cells in each category have a dedicated role to play in keeping one part of the body collective in order. But here's the catch: none of these cells acts independently. All human cells work with one another to keep our bodies healthy and safe.

Take, for example, a paper cut. It's a minor injury, but one that a number of different cell types must work together to heal. As soon as the skin barrier breaks, blood rushes in to clean out the wound and form a clot. This triggers the nerve cells to signal the brain about the problem; in response, the person with the cut feels pain. Immune cells begin the healing process as they wipe out the dead or injured cells, leaving a clean slate. Finally, the skin cells begin to reform in order to bring the area back to normal.

The bacteria living in the human body are grouped into hundreds if not thousands of different species. They arrive from the outside world—everywhere from the soil to the sea.

But what does this all mean for us humans? We don't have all the answers (yet) but we have learned a great deal about the impact of these trillions of microbes on our own health and lives. What I can tell you is this: like human cells, microbes exist not as individuals but as components of an ecosystem. Each of us is an ecological being made up of trillions of cells, both human and microbial. Every cell works as a part of a collective to form what we know is human life. When they all work together in harmony, we tend to be healthy. If, however, the members work against one another, our health suffers. Read on to find out how to get along with your microbes and keep their effect on you as positive as possible.

FRIENDS, FOES, AND BYSTANDERS Every microbial species plays one of three roles in the health of our bodies: a microbe can be friendly, or what is biologically known as *mutual*; it can be a foe, also called a *parasite*; or it can be a neutral *commensal*—from the Latin for "come to the table"—and spend its time taking in the scenery, eating whatever food happens to be around, and not making any trouble.

Our friendly neighbourhood microbes are also known as *symbionts*, because they form a symbiosis in which there is a two-way street of mutual benefit. The pathways are sometimes quite complex, as many symbionts work with specific systems of the human body, such as the immune system, the metabolic system, and even the psychological system. Symbionts are not purposely trying to improve our lives, of course. But over the millennia of human life, we have come to adapt to their presence and benefit from them.

Contrary to their bad rep, only a very small number of microbes are our foes. In fact, fewer than 0.1 percent of all microbial species on earth can cause infection in humans. These foes are officially called *pathobionts*, which is a hybrid of the English word *pathogen* and the Greek word for living, *bioun*. Pathobionts don't make us sick all the time, however. They tend to cause problems only when the time is right for them.

The final and most abundant microbes, the *commensal,* are, as I said, harmless bystanders. They find a home in or on the human body and gorge on the various foods that come their way. In the gut, they chow down on whatever we have consumed. On the skin, they take in all the nutritious goodness that comes in sweat. They also like the mucus in the respiratory tract. Whereas snot is for us a throwaway, for them it's a fine meal.

This general classification—of friend, foe, and bystander—is an excellent way to get a handle on how the microbial population relates to individual health. When a body sample, such as a stool, is taken, there are inevitably hundreds, if not thousands, of species present. Dealing with a list that long can give even a microbiologist a headache. But breaking them down into those three categories helps provide some perspective.

The human body is similar to every other ecosystem on earth. When an environment is healthy, it's because there are more friendly species than foes. The more symbionts we have in our bodies, the better our health. Increase the number of pathobionts, however, and there's going to be trouble.

But there is a twist. Friends sometimes turn on us and become foes. Normally they are happy to be with us, but when the going gets tough, these germs can get ornery. When they do, they can cause us even greater problems than the natural foes do. Having too many friends could therefore be a bad thing. The best option is to have a healthy population of bystanders. They may not do us any good, but they won't cause us any problems either.

TOTALLY MICROBIAL Public health is very prone to using suffixes. Perhaps the most common is *-itis*. Take a noun, add *-itis*, and—boom!—you have an illness. Arthritis, hepatitis, bronchitis, and the eye-cringing keratoconunctivitis are all prime examples. If you're describing an actual event or process, skip the -*itis* and add -*osis*. Halitosis, for example—which translates as "exhaling bad air."

Another suffix growing in popularity is -*ome*, which we use to refer to the totality of a given environment. It was first used in the early 1900s, with *biome*, meaning "all life on earth." It made sense. Then there was the *genome*, which is all the genetic material in a cell. Again, there was little argument. But in the 1980s, -*ome* became a fad. There was *proteome*, for proteins. Want to know about metabolism? You need to study the *metabolome*. How about hormones? You'll need to know the *hormonome*. The list has become so big that there is even an -*ome* for all the -*omes* out there: the *omeome*.

But one -*ome* has stood the test of time. First used over fifty years ago, *microbiome* is a term that means the entirety of microbial species within a given environment. This term is not, however, particularly specific. You can have a building microbiome, a city microbiome, a worldwide microbiome, and, although we haven't yet found microbes in outer space, potentially even a universal microbiome.

Still, the most popular these days—and the one that will be the focus of most of the files in this book—is the human microbiome. Research into the bugs living in and on us has revealed quite a bit about their influence on our lives and health. As you'll see, it's mostly good news.

A MICROBIAL HOME Here's an experiment I have run in the lab. Take a plastic tube, attach nozzles to each end, and then run tap water through it for a few days nonstop. Afterwards, put the tube in an incubator at a temperature similar to that of the human body. After a day or so, something strange appears: a slightly brownish-yellow film of bacteria covering the entire inner surface. It's called a biofilm.

When bacteria enter an enclosed area, like a plastic tube, they tend to spread out everywhere and look for a place to call home. If there are open spaces, they will attach to that surface and begin to grow. If the environment stays friendly, they will spread until they run out of room.

The same thing happens inside you and me. A biofilm can grow on the teeth, or in the throat, sinuses, lungs, stomach, or intestines. The bacteria adhere to a part of the body, usually a nice moist area. As long as there is an open space available, the bugs cling to the surface, and once there, they begin to use up whatever nutrients are available.

It may take only a few hours for a biofilm to start and a few days for it to become established. But once it's in place, it's pretty hard to remove. Even in that plastic tube, I could never get rid of all the bacteria, no matter how much I scraped the area. I would have to use either a strong disinfectant or a large amount of antibiotic.

A biofilm can be either a good thing or a bad thing, depending on the bacteria. Friendly species can help to protect you against infection and also keep nearby human cells safe. But foes will do exactly the opposite. They will attempt to annex even more space—first on the surface, and then, if possible, towards the inside of the body.

The immune system will inevitably react to the invasion by initiating combat. Sadly, without some help (like an antibiotic), the bacterial enemies may never be completely wiped out.

This can have dire consequences. These bad biofilms are responsible for a myriad of troubles, such as tooth decay, cystic fibrosis complications, recurring ear infections, chronic intestinal diseases, and the potential consequences of contaminated needles, catheters and surgical tools.

You may have already experienced a biofilm without knowing it. That slimy feeling on the teeth when you wake up in the morning is an example. This film is new and rather weak, and it can be brushed, flossed, and mouthwashed away. The rest of the biofilms in your body are not so easy to control, and without a daily fecal, sinus, ear, or lung analysis, most won't be detected until it's too late.

Because biofilms start forming almost as soon as we are born, it's important to get friendly microbes in there early (see chapters 5 and 7 for tips on how to do this). If they are the first to arrive, they will take up the space, leaving little for other species. For infants in the first months of life, this is imperative to ensure a proper microbial balance and diversity into childhood and adulthood. The species will change over the years and different biofilms will form as we get older, but a healthy start can keep a baby free from problems when it is most vulnerable.

THE IMPORTANCE OF IMMUNITY Somehow our bodies have figured out a way to allow hundreds, if not thousands, of different species of microbes to live and thrive with us. It's all about tolerance.

Our bodies become used to certain species on and inside us and peacefully allow them to hang around. The decision to accept or reject is up to the immune system. Our immunity is by far the most powerful force inside us. It can affect our metabolism, cardiovascular health, nerve sensations, and even moods. This biological rule of law can be a good thing. When we feel fine, the system is at peace, as are our bodies. But as soon as trouble starts, both need to focus on remedy and healing. The immune system takes charge and commands the other systems to work together to restore a state of corporeal calm.

But there is a down side. If the immune system is continually in hyper-drive, the other systems may not work optimally. They may even lose the ability to keep us healthy. Remember the last time you had a cold or the flu? Those sniffles, aches, pains, fever, and bouts of nausea and diarrhea were symptoms of an overtaxed immune system. For sufferers of chronic diseases such as allergies and asthma, this dark side of the immune system can be a real nightmare.

This is where tolerance comes into play. When a strange bacterium, virus, fungus, or other microbial species shows up in or on the body, the immune system assesses it—friend, foe or bystander?—in the cellular equivalent of a border crossing. Immune cells examine the visitor carefully to determine its traits. In the meantime, the immune system's population in the area stands at the ready in case a threat is identified. After analyzing all the components of the visitor, the system reaches a decision. If the microbe is found to be a friend or bystander, the immune cells allow it to move on without incident. If deemed a foe, the visitor is dealt with quickly, usually with summary execution. Any other microbes from the same species that happen to be in the area will also fall victim to attack.

Over time, as microbes continue to be inspected, the number of tolerated species increases. The immune cells become used to the presence of these microbial friends and continue to keep them around. But this doesn't mean the situation cannot change. To maintain approval, the microbes have to be recognizable over the course of a person's lifetime. Microbes are known to evolve, and may change their appearance and their activities. When they do, tolerance may be lost and the friend or bystander may once again have to be judged safe or be exiled or killed.

There's another way in which tolerance is important to our health. By keeping our defences from attacking every single foreign entity that comes along, we conserve energy and avoid undue consequences. For example, when engaged in highly active combat, the immune system forces our metabolism to change, diverting energy from our normal cellular activities to one that supports a strong defence. This can quickly cause us to feel drained and lethargic. Tolerance ensures this effect occurs only when needed.

But the loss of vigour isn't the worst problem associated with an active immune response. The soldiers have the ability to kill human cells as well as microbes and can wage a war in which killing is indiscriminate. This is not an entirely bad thing. When a cell is infected with a virus, the immune system cannot reach the hidden entity. So it just kills the cell instead. When this happens on a small scale, there is no need to worry, but when tolerance isn't in place, this particular tactic can kill large areas of tissue as well as organs. Much like an allergic response, the inability to tolerate certain exposures to allergens, toxins, and pathogens can be life-threatening. One form of this kind of widespread damage is acute respiratory distress syndrome, or ARDS, which can eventually cause the lungs to shut down.

There are, of course, less deadly problems associated with intolerance to microbes, such as dermatitis, sinusitis, and rheumatoid arthritis. Regardless of the threats to health, all these conditions can be prevented with proper immune tolerance.

Achieving immune tolerance is a lifelong task and requires constant contact with a highly diverse microbial population. This ensures that the immune system is as prepared as possible for any exposure. To accomplish this, there is no better place than the great outdoors, particularly rural areas. In urban areas, the diversity of the microbial population tends to go down, shifting towards those species usually found on and in the human body, including many pathogens. In areas with a low human population, we are exposed to a wider assortment of microbes. It's our best chance to find our friends and, more importantly, to keep our defence forces focused on foes.

OUR MAGNETIC SHIELD Our intestines—an incredibly long tube of tissue—are all that stands between our food and drink and the rest of our insides as we digest our meals. When our bodies have a temporary need for a little extra of something— glucose, say—the cells in our intestines have mechanisms to extract it from the mix. When there is just a general need for a faster flow of nutrition, the cells can loosen up and make the barrier a little more permeable.

Gut permeability is a highly controlled process and only happens when there is a real need for it. But, if the cells are damaged or start dying in large numbers, the intestinal structure weakens and small holes open up. For unfriendly bacteria wishing to cause harm, it's an invitation to invade.

The cells have a secondary shield to protect against invasion in these areas. But this one isn't physical, it's electric. In essence, the intestines prevent an attack by forming an electrified fence around the weakened cells, protecting them until they can return to normal.

Here's how it works. Viewed under the microscope, intestines are not smooth but have very tiny hair-like extensions called microvilli. These generate electric power by way of biological magnetism. Positive and negative charges are separated from each other to form poles. The microvilli ensure that the negative pole is always on the outside of the cell. This is important because bacteria, particularly those that are out to harm us, also keep the negative charge on the outside of the cell. When the bad bugs come near the intestinal cells, they are repelled. It's like trying to push together the same poles of two magnets. This is an excellent defence mechanism because it's local and does not involve the rest of the immune system. This way, the cells are kept safe and the body is none the wiser.

LATIN FOR GERMS Latin is a part of scientific life. Scientists don't converse in it, but they always seem to be using it. It's a universal way to recognize a species of organism. It doesn't matter if we're talking about an apple (*Malus domestica*) or dog

(*Canis familiaris*) or even you and me (*Homo sapiens*)—when we need a scientific term for living organisms, we are likely to turn to Latin.

Most microbial species have Latin names. Some reflect traits, like *Clostridium difficile* (a pathogen), which essentially is Latin for "spindles that are hard to grow." Sometimes, the name is derived from the person who discovered the species, like *Escherichia coli* (Latin for "Escherich's fecal bacterium"). Then there are some with names that obviously try too hard to be descriptive, like *Desulfovibrio dechloracetivorans*, which is Latin for "sulphur-loving wriggly bacteria that uses chlorine and acetate."

Over the last half century, researchers have attempted to simplify matters by using names derived from modern English, particularly for viruses. Take the Heartland virus, named for its geographical origins in the United States. Or HIV, which stands for human immunodeficiency virus.

But a name, it seems, can be misleading in any language. A virus commonly referred to as MERS stands for Middle Eastern respiratory syndrome. Unfortunately, this disease is not restricted to the Gulf States and has spread to other areas of the world. Some viruses are given Greek symbols rather than names, such as the λ and φX174 bacteriophages. Then there are names that could have used a little more thought before they were given. For example, about half of us carry a particular virus in our guts and excrete it in faeces. It's called crAssphage. And no, it isn't named after its source, but after the type of computer program—a cross-assembler—that helped discover it.

TAPPING THE KEGG The human body is designed to make access difficult, so spying on bacteria in various bodily areas such as the sinuses, lungs, and gut is no easy task. The best we can do is wait until the bugs come out of us and then analyze their history.

Tracing a bug even retrospectively is a complicated business. The bacterium has to be quickly analyzed at the genetic level to determine which activities it performed. This in itself can be a frustrating task, as a bacterium can perform dozens, if not hundreds, of tasks simultaneously. Some of those tasks are related to each other, while others are independent. Sorting through that information is both time-consuming and mentally challenging. Thankfully, there is an electronic tool to help sort everything out. Since 1995, the Kyoto Encyclopedia of Genes and Genomes—or KEGG—has helped researchers worldwide understand exactly what a bacterium has been up to inside our most unreachable places.

At its core, KEGG is a database providing details on certain biological functions, such as eating, multiplying, or making toxins. Within each function is a collection of genes, all of which are needed to accomplish that particular function. For a researcher, KEGG is a fantastic tool to help wade through the information and determine what the bacteria have accomplished. All scientists need do is log into the KEGG website, enter the names of the genes found, and wait for their functions to be revealed.

In the context of health, the true value of KEGG comes in determining the relationship between bacteria and disease. In some cases, there may be no link at all, while in others distinct

patterns can be seen, suggesting some bacteria are indeed act-
ing as foes. Although KEGG cannot offer answers on how to
resolve a situation, it can provide a reference point to assess
medical options in the lab. This can increase the information
learned long before a treatment is ever tested in the human
body and so reduce the time needed to reach clinical trials.

DEADLY DOUBLE-CROSSERS Microbes more often than not
work together to keep us healthy. But sometimes a friendly
species turns into our worst enemy. Many life-threatening conditions—flesh-eating disease and meningitis among them—
are caused by these double-crossers, and many lives have been
lost as a result of our not catching on to them soon enough.

Figuring out why a friend becomes a foe is a real challenge. When a bacterium turns on us, it's usually because of
something we did (probably inadvertently) or something that
occurred in our bodily systems (probably in order to keep us
alive). It all comes down to stress. Bacteria hate a stressful
environment, and when they encounter one, they choose one of
three paths to deal with it: fight, flight, or sporulate.

Only a handful of species are able to sporulate, which means
to go into a form of hibernation until better conditions arrive.
Those that choose flight simply hitch a ride with whatever
fluid happens to be flowing out of the area at the time. If
bacteria choose to fight, however, a battle with our protective
force, the immune system, may ensue.

Bacteria choosing to fight have some effective strategies.
Their weapons of choice are toxins, and many bacteria have
an entire arsenal of them at their command. In flesh-eating

disease, toxins are used to break down the cells of the skin, causing significant damage. Some toxins may evade the immune system, causing harm outside its reach. The bacteria known to cause meningitis are experts at this, as they can find a way to the brain, where they lead to cellular destruction and swelling. Others attempt to deflect the defences of the immune system and shift it to a self-destructive path. The bacteria that cause pneumonia, for example, mimic certain cells of the body so effectively that the immune system attacks itself. In the meantime, the microbes grow and multiply and worsen the condition.

We have a clear understanding of how these bacteria react with each other when under the influence of stress, but we know rather less about what triggers that reaction. What we do know is that a change in our own bodily function, such as a fever, can be enough to transform a microbe from coexisting friend to dangerous foe.

One of the best-studied examples of this is bacterial pneumonia during an influenza infection (it occurs regularly and was the cause of most deaths in the 1918 Spanish flu pandemic). Normally, these killer bugs are happy members of the bacteria in our noses. They enjoy a good life there and return the favour by helping to crowd out hostile species. But all this mutual benefit changes when the body begins to fight influenza. As part of our response, we turn feverish and the inside of the nose begins to warm up. The bacteria sense that there's trouble ahead. Instead of waiting for the killer blow, some decide it's best to move on—the flight response—and look for another, perhaps more hospitable, place to reside. They loosen up from the sinuses and are inhaled into the lungs.

Unfortunately, the conditions there are no better; the lungs are hot and sticky and full of immune cells trying to fight off the virus. When those immune cells come into contact with the bacteria, it's like getting slapped in the face with a dueller's glove. The bacteria realize that running away again isn't the best idea; it's time to fight.

This microbial battle begins with an attack. The bugs release toxins to kill any nearby lung cells. Next, the new enemy employs a defensive posture, surrounding its fighters with a thick shield called a capsule. This protects against any immune counter-attack. With the human body in distress, the bacteria then go for the kill shot, releasing chemicals such as hydrogen peroxide to kill immune cells and anything else that might be living nearby. Within a matter of days, this microscopic battle can have macroscopic consequences for the patient, who may be left in a life-or-death situation.

The actions of the bacteria behind pneumonia may seem extreme in light of the relatively mild stressor of fever. Yet this is just one example of how a single change, however slight, can lead to severe consequences. For this reason, as we learn more about the microbes inside us and how they interact with us, we gain more insight into how a friend turns into an enemy and how we might prevent this from happening.

OVERKILL The immune system isn't entirely trustworthy. Sometimes when its troops attempt to kill a real or supposed enemy, they go into an uncontrolled rampage, resulting in damage to our bodies. If the region affected is a vital organ such as the liver or the pancreas, those rampaging forces could end up killing it—and us—in the process.

Thankfully, we have a group of cells whose sole purpose is to rein in the troops and keep them from causing undue damage. They are known as regulatory cells. They work 24/7 from the moment we are born to monitor the immune system and ensure that any attacks are solely against threats and not against us.

The process is quite simple. A regulatory cell enters a combat zone and determines what exactly is being attacked. If the target is part of our own body or a friendly microbe, the regulatory cell sends out signals to stop the campaign and force the defences into a state of calm. The troops obey the command and slow down the attack. Eventually peace is returned and the regulatory cells head off in search of another battle.

What's great about regulatory cells is their memory. Once a friend is recognized, that information is stored for as long as we live. This has two benefits. First, it keeps our immune system focused on known foes. Second, and most important, it ensures we don't fall victim to an unnecessary internal attack.

EARLY WARNING We are covered in about two square metres of skin, but that is not the body's biggest surface area. The respiratory tract is 160 square metres and the intestines over 250 square metres. All that acreage is vulnerable to attack from the outside world. And our insides, of course, are at greater risk of a surprise attack than our outsides, which we can keep an eye on.

So it's just as well that our bodies are able to look out for potential foes without our conscious assistance. Each one of the billions of cells lining the eyes, skin, sinuses, lungs, and intestines carries a specialized set of proteins designed to determine if a visitor is friend or foe.

One of the most hard-working of these proteins is called a toll-like receptor, or TLR. It is one of the sentries of the body's defence systems, prompting chemical messages inside a cell. If the TLR senses a toxin, a pathogen, or some other possible invader (such as an allergen), it immediately warns the rest of the body of a potential attack. If, however, there is no threat, the forces are instructed to stay at ease.

The value of the TLR is twofold. First, the early-warning system provides the means to stay on top of every exposure over a massive area. But more importantly, the TLR is an energy-saver. By keeping the body's defences calm when not needed, the TLR avoids wasting vital resources on futile battles.

ME AND MY MHC Take a random sample of human beings and identify their microbes. On average, these people will have only about 30 percent of the same types. Those with a considerably larger number of shared microbes are usually related. Twins and siblings may share up to 60 percent of the same microbes, and mothers and their children around 50 percent.

It's hardly surprising that members of the same family have similar microbial populations. After all, they live together, eat the same foods, and participate in similar activities. They should have the same microbes. But there is another factor involved, and it has nothing to do with social interaction.

There is on the surface of most human cells a protein called the major histocompatibility complex (MHC). The MHC has one job: to screen foreign visitors to determine if they are friend or foe. But before the MHC can accomplish its task, the unknown entity must first be processed for assessment.

When a microbe is first encountered by the human body, it may be allowed access to a cell. But this isn't always a friendly welcome—the visitor may end up being killed. This act of microbial murder actually has a purpose, however. It's called antigen presentation and serves as an early warning alert for the immune system.

Here's how it works. When the microbe enters the cell, it will be immediately attacked by acids, peroxide, and other lethal chemicals, killing it. The corpse is then broken down by a group of enzymes forming several small pieces. These are known collectively as antigens and act as microbial fingerprints. The MHC grabs these fingerprints and offers them up to immune cells for a check on whether the microbe should be tolerated. The immune system has memory of previous encounters with antigens and can quickly make a decision. And once that decision is made, any future appearance of the same kind of microbe will spark the directed response.

The number of antigens in the world is countless, so the immune system can create an incredible diversity of MHC types. It's conceivable, mathematically at least, that each of us could have up to a quintillion (that's a one followed by eighteen zeroes) types of MHC molecules. In reality, we have far fewer, usually between twelve and eighteen. Evolution has deemed that this limited number is sufficient to deal with the thousands of antigens we encounter in our lifetimes.

Heredity decides which twelve or so MHC molecules we will carry with us through life. A mother's genetic makeup, or genotype, determines which ones will be passed on to her child. This inheritance is beneficial to the infant, who has so far developed little immunity and requires help from the mother through continual contact with her and with her breast milk. The MHC can detect the goodness of not only the nutrition in the breast milk but also the microbes contained within. This improves the chances for good microbial settlement in the first few months of life.

As the child grows into an adult, the MHC helps to determine which microbes will be able to stick around throughout the course of life. The child's overall bacterial population continues to be highly diverse and will differ from the mother's over time. But that core group of MHC-tolerated microbes—up to 50 percent of them—will always be welcomed as friends.

IDENTICALLY DIFFERENT Identical twins—those apparent outliers in the golden rule of human uniqueness—have captivated scientists for years. And not because of their similarities, but for what sets them apart. When twins are observed up close, their differences become strikingly clear. Their fingerprints don't match. They have individual psychological traits. Even their genetic material has subtle variations.

For microbiologists, twins are the best subjects to help understand our relationship with our microbial population. By the first month of life, identical twins have a bacterial profile that differs by as much as 40 percent. They are kept in the same house, fed by the same mother, and have the same activities. Yet they cannot keep identical types of microbes in their gut.

This anomaly makes for a great platform for investigation, and many studies involve examining identical twins' microbes and associating them with various aspects of their lives. The process is simple: take twins, identify the differences between them—say lean vs. obese—and then look at the bacteria inside each one. Do this enough times with enough pairs of twins and a pattern begins to emerge. So too does the ability to determine a friend from a foe. Thanks to these studies, we know quite a bit about how microbes affect each of us.

Each day, the number and types of bacteria inside and on us change. New bacteria can be brought on board in any number of ways—diet, exercise, travel, infection . . . With each experience, bacteria visiting the body can become established there. Sometimes this is good, but other times it can lead to problems, as foes take advantage of certain human actions and cause trouble.

What becomes clear is that most of the time, microbes do not control us—we control them. But if we ignore them, we may end up allowing them to engage in a coup and change our lives for the worse. The path to better health may lie in continuing to keep our friends close and our enemies in line.

DIVERSITY IS THE KEY TO HEALTH Good health depends on having a diverse bacterial population made up mostly of harmless bystanders (the commensals), a good number of friends (the mutuals, or symbionts), and few to no foes (the pathobionts). This kind of variety creates a buffer against infection and leaves few available spaces for microbial attack. Unless the invader has a specialized means to cause problems, there is simply no opportunity available. And there's an added bonus: many of our friends and some bystanders have the ability to fight off infection-causing troublemakers using their own antimicrobial weaponry. It's a double blow to the pathobionts, making them even less likely to cause trouble.

One of the best-known examples of a health problem caused by a lack of diversity involves the nefarious bacterium *Clostridium difficile*, or *C. difficile*. It's made a name for itself as a rather nasty pathogen. It targets the interior lining of the gut and leaves a person at risk of diarrhea, cramping, pain, and internal bleeding. Over the long term, the bacterium can erode away the cells of the gastrointestinal tract and enter the body, leading to weight loss, kidney failure, and eventually death.

Every year, *C. difficile* kills tens of thousands of people. But some 5 to 10 percent of the human population can accommodate *C. difficile* without experiencing any difficulties at all. The difference between infection and coexistence comes down to one factor: microbial diversity in the gastrointestinal tract.

When *C. difficile* enters the body, it heads to the colon. If there is a significant amount of diversity, the number of nutrients for the pathogen is limited. So too is the space for attachment to the cellular lining; that space is already occupied. But when that diversity is low, there is ample room for attachment and also an excess of available nutrients, making it easy to start a colony. The pathogen takes full advantage of this opportunity. As it grows, it takes up more and more space and gobbles away the nutrients, leaving less for other bacteria.

If allowed to progress, the problem can get much worse. As the nutrients run out, the cells get hungry and begin to raid the host body's supply. A toxin is produced to kill human cells and release their valuable food. Eventually, the harm spreads from the gut to the rest of the body as the toxin, and eventually the bacteria, gains access to the bloodstream. By this point, the fight is already lost because the immune system is helpless. It can only stand by and watch.

Cultivating a diverse microbial population in the gut is not all that difficult. Just eat natural foods and avoid processed varieties! Okay, so maybe it's not that simple for most of us. But we should try to get raw vegetables and fruit in our diet, along with fermented foods and probiotics. They help the friendly species in the gut to thrive.

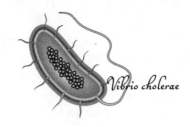

Vibrio cholerae

2. THE ENVIRONMENT

LOCATION, LOCATION, LOCATION Where do our microbes come from? Some transfer to us from other people, but most are from the soil, the air, and the water. No two environments in the world have the same microbial population, and the humans who live within those environments will be exposed differently. Also, because of their different ways of life they take on different kinds of full-time resident. Take a person from the United States, another from the African country of Malawi and a third from the Amazonian area of Venezuela. For the first year of life, these people all have in their guts similar microbial species usually attributed to bacteria known to love breast milk. But after their first birthdays, the microbial populations for each change dramatically. By three years of age, these people are so far apart, microbially speaking, a biologist could determine where they live simply by looking at their feces.

We can influence our overall exposure to microbes by seeking out greater diversity. We can take a trip out of town (see "Air Freshener," below). We can eat more raw fruit and vegetables, which will deliver a number of good, healthy species straight to the gut, where they can increase diversity and reduce the chances for pathogen invasion. Another option is to take probiotics (see chapter 5).

AIR FRESHENER As soon as I am about fifty kilometres from the centre of the city where I live, I take a deep breath and something in my brain clicks. Suddenly, everything feels better. Most of this change is psychological, but microbes play a role in the switch from city insanity to country calm. It has to do with the diversity of the microbial population not inside but around us.

The average urban environment is essentially a petri dish of human microbes. Although you will find certain species associated with water and soil, including those that cause wood to rot, most of the bacteria in urban settings comes from human skin. Now, compare that lack of diversity to, say, a forest or rural area like a farm. Here, the diversity is huge. Bacteria mingle with other organisms, including fungi, amoebae, and even worms. The population is also stable, so there are no significant shifts over the course of a day or a year. These microbes also help to keep the air fresher by reducing the levels of microbial toxins. Every breath is a clean one (though it may be a little smelly if you happen to be near a farm).

Although increasing our body's diversity by getting out of the city may bring a general sense of well-being, it is short-lived. Almost all species encountered in rural areas won't stay with you for long. Eventually, you will go back to the usual microbial grind. But this isn't entirely a bad thing. It simply means you'll have to get out of the city more often.

AVOIDING ARSENIC Arsenic is everywhere. And because arsenic compounds dissolve easily in water, foods that are grown in water—rice in particular—are at risk of contamination. You've probably been exposed to arsenic yourself through what comes out of the tap. Thankfully, arsenic is present in drinking water in extremely small quantities that our intestines can deal with easily, shuttling it out of the body so we are kept safe. The rest is taken to the liver, where it is targeted for excretion and dumped in our urine.

But this process isn't perfect, and some arsenic may stick around. With continued exposure, we accumulate more of the element and increase our risk of being poisoned.

But there is another problem. The microbes in our gut also are not fans of arsenic, and they have a very different—and less effective—way of removing it. Instead of trapping it, they modify the chemical's properties so it can be quickly and irreversibly removed. Unfortunately for us, this modified form of arsenic, while safer for our microbes, is actually *more* toxic for us, as it can pass through the gastrointestinal barrier and make its way into our bodies. This modified form cannot be metabolized and ends up accumulating, which leads to an increased risk of cancer, cardiovascular disease, and diabetes.

There are ways to reduce exposure to arsenic. Water filters can trap the chemical and help to improve water quality.

Some are highly specialized, while others, like sand filters, can be made at home. Also, if you like to eat rice, you might want to make sure it is arsenic-free. This may not be easy to determine, as arsenic is not listed on food labels. But several consumer and government organizations have developed lists of arsenic-free products. Although the amount of arsenic to which any of us is usually exposed is incredibly low, zero is always best.

WINDOWS OPERATING SYSTEM My home, your home, and indeed all indoor environments are teeming with bacteria, viruses, and fungi. Not to mention insects. These cohabitants are all part of a living ecosystem—an indoor biome. Depending on the nature of these inhabitants, the home can be a happy one or a nightmare.

The difference between a healthy home and a harmful one can really be brought down to a single facet of modern engineering: it's called heating, ventilation, and air conditioning, better known by the abbreviation HVAC. Most modern homes and buildings have some kind of HVAC system to ensure that air is not only kept at a certain temperature but also circulates to avoid staleness. This presents a problem for our microbial health. With the assistance of an HVAC system, bacteria can easily travel through the air and spread all over the entire building. As they travel, they may come across a nice warm, moist environment where they can start making biofilms. There appears to be no end to the potential for a biofilm, as bacteria can be found on everything from doorknobs to light switches. The most common places to find them are the bathroom and the bedroom, with the kitchen coming in close behind.

Microbes are also spread through the environment through a process called shedding. Much like a pet leaves hairs everywhere it goes, we leave behind traces in the form of bacteria. When we change our clothes, move from one place to another, engage in play and retire to bed, we are constantly shedding. As we do, the HVAC system picks up the bacteria and spreads them all over the home, including into areas we don't think about, such as the ducts and vents. Over time, the diversity of the building's microbial population will skew so that it contains mostly human bacteria.

When someone who's ill enters a home, the diversity shifts. Instead of being rich with a variety of microbes, the environment becomes overwhelmed with the pathogen as the sick person sheds on surfaces, in the air, and on other people in the vicinity. The immediate spread of infection is not the only concern. Even after the person has left the area, certain bacteria can persist in the environment.

When a pathogen is contained within an enclosed environment, it can sporadically cause infections over several years. One of the longest-lasting pathogens has to be a species known as methicillin-resistant *Staphylococcus aureus* (MRSA), which causes skin and respiratory infections. It's usually thought to be limited to healthcare facilities, but this microbe can enter the home and has been known to stick around for as long as eight years. All the while, the household members might end up with a variety of health problems.

There is a way to prevent problems associated with HVAC systems and reduced microbial diversity. It's called Microbial Dispersal Capacity Maximization. The process involves using a controllable vent to the outside world to increase the diversity of the bacterial population beyond human, pet, and brick microbes.

It's also known as opening the windows.

Once fresh air gets inside, the effects are almost immediate. The new bacteria find places to settle, and then seek out and kill any lurking pathogens with their antimicrobial peptides and other antibiotic-like chemicals. In a matter of as little as a few hours, the air is returned to a more natural state for all to enjoy.

At one time, this intervention was easy, but many buildings today have sealed windows, leaving those inside fully dependent on HVAC systems. Some newer buildings have incorporated vents in place of open windows to allow fresh air to enter the system. It improves diversity and keeps the air safe to breathe. For anyone with no access to a workable window or fresh-air HVACs, the best option is simply to go outside from time to time. While the weather might be frightful, the effect of having a change of diversity will be delightful. Another option is to bring nature home. Plants are an excellent way to increase microbial diversity and can also provide some welcome aesthetic relief to modern-day architecture.

A BETTER MASK There are times when we simply cannot avoid sick people. For me, the worst is being cooped up with hundreds of other people on an airplane. Healthcare workers are encouraged to use masks to protect themselves from illness (or to avoid passing illness to others, if they are the sick ones).

But outside of the hospitals, clinics, and doctor's offices, the sight of a blue mask is not quite as accepted. People wearing them tend either to be viewed as germophobes or shunned as ill themselves. (Actually those mask-wearers are quite often—and quite sensibly—protecting themselves from pollen.)

I recommend a mask substitute that is a simple and fashionable piece of clothing we can all add to our everyday wear: a scarf. When it comes to preventing a pathogen from getting into your respiratory tract, a scarf does a fantastic job. It contains a lot of absorbent material with a high surface area. It's designed to help retain heat but can also trap even the smallest amount of liquid. When a sick person coughs or sneezes in your vicinity, the scarf acts as a barrier, keeping infectious droplets of mucus or saliva from getting into your nose or mouth. It's simple, effective, and inconspicuous—or at least less conspicuous than a mask. The only caution is to ensure that the scarf is not tightly wound around a child's face. Although preventing bacteria and viruses from entering the respiratory tract is laudable, stopping air from getting into the lungs is not.

MOBILE MICROBES I'm often asked to name the germiest place in the home. I have two answers. If you're looking only for pathogens, then the bathroom and kitchen sinks take the prize. But for the most microbes in the smallest area possible, the clear winner is any mobile device.

Take a look at the screen of your cellphone or tablet. Unless you've cleaned it lately, you'll see both smudges and smears. Chemically, this is a combination of oil and dirt from your skin and dust from the environment. Within that mix are microbes—sometimes thousands of them. They are left here every time the device touches your hands or a surface.

But don't fret. Unlike the sink, your mobile device mostly carries harmless bacterial species. There may be a few fecal bacteria, especially if hand hygiene isn't the best or the device is used while conducting bathroom business. But the likelihood of infection is fairly low. If you are in good health, you would need to inhale or ingest an entire screen's worth of bacteria—and even then, it may not be enough to cause trouble. An inadvertent finger lick after touching the screen offers no risk.

But if you're talking viruses, such as those that cause colds and the flu, it's a different story. A far smaller number—say one to one hundred—is required to cause an infection. You could easily transfer this amount by licking your fingers after touching a contaminated screen. For this reason alone, it's best never to share your phone with a sick person or leave it out in the open around coughers and sneezers.

Even though the risks from catching a mobile-mediated infection are quite low, it's still prudent to give the device a good sanitizing about once a week. Alcohol wipes are best, as they kill bacteria and viruses, dissolve the oil, and leave no annoying streaks. Eyeglass cleaners are also quite effective. Just don't wash the device in water. Not only will that fail to remove any of the contamination, but it might end up destroying it altogether.

PUT A LID ON IT Do you close the toilet lid before you flush? Some people seem to forget there are two covers for the toilet (well, at home at least). Others think the lid is there mainly for aesthetic reasons, or to provide a seat or keep a pet from drinking the water.

But there is a greater purpose to the lid: it keeps the bacteria in the bowl. When you flush, the rush of water creates a flurry of aerosols that can travel up to two metres from the surface of the toilet water. What's in those droplets? Well, it's whatever might have been in the bowl at the time. For an average bowel movement, this could mean hundreds to thousands of bacteria and viruses. If it's watery, the number can be even higher. Normally, there aren't enough bacteria to cause infection, but in some cases—say, a virus that causes diarrhea and vomiting—one open-lid flush can be enough to spread the disease.

Here's where it gets a little troublesome. Those droplets can end up on surfaces like the sink, the taps, the floor, and yes, the toothbrush. Does this mean a lidless toilet dooms us to illness? Not really, because most of us have the proper immunity to keep us healthy. Yet in places such as hospitals, where people are far more susceptible to infections, this accumulation of bacteria and viruses can be troublesome. Without regular cleaning with disinfectants, a toilet stall may become the source of an outbreak. When you're at home, you should get in the habit of closing the lid every time you flush. If you're in a place where there is no lid, like a public bathroom, just hold your breath and get out of the stall as soon as possible. If you happen to be following someone, give it a good thirty seconds before you head in. The droplets will settle, and the air, albeit stinky, will be safe to breathe.

NOW WASH YOUR HANDS Each visitor to a public lavatory sheds millions, if not billions, of bacteria while attending to nature's call. But these unavoidable places are not quite the germophobe's nightmare you might think. Most of the microbial species found here have little to do with disease. Some pathogens are present, having been deposited by sick individuals who choose to head out rather than heal at home. Thankfully, their contribution to the overall population is minimal. Even more comforting, these pathogens tend to lurk in places unlikely to be touched by others—most commonly the sink, where they end up after handwashing.

Pathogen pickup zones include the toilet and faucet handles, the doorknobs, and the soap dispenser, but even here there is no real threat. The actual number of bacteria present in any one of these spots tends to be in the hundreds. This is nowhere near the amount needed for most bathroom bacteria to cause infection. Our immune systems and the diverse bacteria already inside our bodies see to that. The more species we have, the more effort a pathogen needs to get a hold in the body. It takes quite a few pathogens to cause problems; at least one hundred thousand would have to be ingested to make us have gastrointestinal distress.

Even if an infected visitor to a bathroom expels that number of pathogens, most of them will end up going down the drain. Some will get caught up in the air currents of a flush and make their way into the atmosphere. But their final resting place will most likely be the floor, not another person's mouth.

Once a resting place is found, though, the troubles for the bacterium are not over. It will have to find nutrients. This may be a problem, as typically the only possible food sources are the biological wastes left behind by human visitors on toilet seats, on handles and knobs, and in the sink. There may be organic matter on the floor, but the supply is usually insufficient for optimal nutrition.

If the pathogen can successfully jump through these hoops, there remains one more hurdle: the other microbes in the area. Most public bathrooms settle into having a nicely diverse microbial population. (Regular cleaning lowers the total number of bacteria, but not their diversity.) This happily balanced group, unenthused about the new visitor, may decide to attack. When this occurs, the chances for the pathogen are next to nil. Only a few species have the ability to withstand this stage of attack and maintain a presence in the area. The rest are destined to be eliminated.

But public washrooms are of course not risk-free. As the number of people coming and going over the course of a day increases, so will the level of microbes in the air and on surfaces. Depending on the hygiene of each visitor, the quantities of bacteria left behind could be enough to cause infection. Though the bathroom ecology will eventually return to normal, there is a temporary threat lurking there. As such, the best way to ensure protection is to seek out soap and water for the hands.

AGENTS OF CLEAN There used to be just one disinfectant capable of killing germs: bleach. Today, store aisles are filled with an endless variety of cleaning brands, each with an appealing name and an eye-catching label.

Despite the diversity of these products, there are only a few actual disinfecting chemicals out there. The most common is still bleach, and it is still considered the gold standard. Soap is also quite popular, although in these formulations, the molecule looks quite different from the one used on the skin. (It's been modified to specifically kill bacteria and viruses.) Toxic chemicals like ammonia are also a favourite. They are used in low concentrations to kill the bugs but not harm us.

Hydrogen peroxide has been used for over a century to help clean wounds. You've probably seen it in a brown bottle at the drugstore. But that type is specially made for the skin, meaning it doesn't have the strength needed for general cleaning. Peroxide products have the advantage over bleach, which can corrode metal, of being safe to use on any household surface.

Most natural alternatives contain antimicrobial oils from plants or ores. The first such extract was from coal, and it's still used today in several products. It's since been joined by pine, tea tree, and citrus extracts. Though these less synthetic versions may offer green-minded folks peace of mind, they are usually less effective at killing bacteria, viruses, and fungi.

A more natural and effective disinfectant is steam. It eliminates germs in a matter of seconds and is easy to make—you just boil water—but getting it where you need it can mean adding a hose or some other complicated apparatus. Steamers are now sold almost everywhere, and they range in size from small handheld devices to large vacuum-cleaner styles.

DOG LOVERS UNITE Human couples enjoy a measure of microbial harmony, but dogs and their owners can have an even greater number of shared bacterial species. That is, on the skin at least. What's even weirder is that two human strangers owning dogs can have more similarity in their skin's microbial populations than a cohabitating couple. This trait is unique to dog owners, as no other species has the ability to change a human's microbial population to the same extent.

There are two major reasons for this incredible bacterial bond. First, because dogs enjoy getting acquainted with a more varied population—otherwise known as rolling in the mud—they have a higher microbial diversity on their skin. When they head home, they bring those species with them and share them with the owners. Second, most dogs have lots of hair, and microbes can live on every millimetre of it. As dogs move around the house, they shed. When they are petted, they shed, and when they are kissed, they shed. The bacteria released into the air through our interactions with our pets end up on our skin, in our mouths, and sometimes in our guts. Dogs and humans have adapted to harbour simi-lar microbial species, which means any species introduced into the family environment will be greeted warmly rather than treated as enemies.

FIDO VS. FLUFFY Dog lovers, they say, tend to be outgo-ing and dominant, while cat lovers are more introverted and unaggressive. If we extended that premise to microbes, we would expect a dog to be more likely to host harmful species while a cat would have healthy ones. This is, however, not the case, as most microbes—friends and foes—find dogs and cats equally appealing.

These two mammals have similar external and internal environments, and so they attract similar species. But there are differences. Cats have a greater diversity of species in the gut, and they carry a greater number of bacteria known to be human foes. They also carry non-bacterial species such as yeasts, fungi, and protozoa. In comparison, dogs are more akin to humans, with a higher percentage of friendly bacteria. Also, you are far less likely to acquire an unwanted microbe from a dog's feces, as they do their business outdoors. Count one for the canines.

Cats are continually cleaning themselves, making their microbial population fairly stable. Dogs, on the other hand, are always out and about, collecting a variety of species to bring back home. Although exposure to diverse microbes is usually a good thing, this could increase the potential for tracking in infectious diseases. Advantage, felines.

Then there is the mouth, which is perhaps the most important part of the equation. This is an important route of microbial transfer, as dogs and cats share their saliva with their owners constantly. But in this case, dogs and cats are tied. They both have diverse bacterial populations in their mouths. Most are harmless to us, but some are pathogens that can cause skin and possibly blood infections through a nasty bite or scratch.

Tallying up the advantages and disadvantages, there is no clear winner between cats and dogs, at least microbiologically. They both share their bugs with us and pose little risk of infection.

Whether you own a dog or a cat, careful hygiene is the way to avoid trouble from pathogens. This means handwashing for us, regular baths for the dog, and routine disposal of litter for the cat. Infants under six months old, and therefore with weak immunity, can safely be in the vicinity of pets but should be prevented from kissing and cuddling them.

CAN BACTERIA SAVE THE BEES? We need bees for human health—not just because they make honey. Their most important contribution comes in the form of pollination. Without them, we could lose significant percentages of crops such as canola, blueberries, tomatoes, and almonds, to name only a few.

But bees are having a rough time these days. In a normal year, populations can decrease by as much as 15 to 20 percent. But recently some colonies have suffered losses of up to 90 percent. At that rate, the bees cannot recover, and they will eventually disappear. This crisis has a trifecta of causes. Climate change is shrinking the bees' habitat. Pesticide use causes chronic problems in their metabolism and neurological function. And perhaps worst of all, bees are under attack from several microbial invaders capable of collapsing a hive and killing the thousands living within.

There's little we can do about climate change in the short term. Controlling pathogens is most certainly not a viable option. As for pesticides, there's a raging debate, with farmers arguing that reduced use will result in plummeting crop yields. But there may be one way to improve the bees' chances for survival: we can incorporate good bacteria into the mix to improve their ability to fight off infections in their environment. Certain bacterial species can improve bees' immunity. When these are provided as a supplement to nectar or honey, the insects' internal defence forces grow stronger. When these bacteria are administered to a hive, bee health improves and losses are reduced.

Though this sounds like a great way to save the bees, there is a major problem: getting these bacteria to all the hives around the world is simply impossible. To succeed at this, we have to bring the bees closer to us. This entails, first of all, finding the most vulnerable populations and then transferring them to safe havens such as bee hotels or rehabilitation centres. Here, they can re-establish their population before returning to the wild. But once out there, the same issues can recur, once again threatening the population. This leads to a vicious cycle of rehab and relapse.

For the moment, this Band-Aid solution appears to be the only one available. Yet as we continue to deal with the larger issues of climate change and pesticide use, longer-lasting answers may come to light. Until then, we may have to rely on our microbial friends to help sustain the bee population and ensure crop security is maintained.

THE VALUE AND RISKS OF COMPOSTING We all know that composting is good for the environment and the garden. But is it good for your health? The answer depends on the size of the composter as well as what is thrown into it.

Backyard composting—the most common form—can reduce most household waste by up to 80 percent. The fruit and vegetable throwaways are useless in a landfill, but when piled in a composter with some soil, they can turn into a wonderful collection of nutrients for plants. When only these two types of foods are used, the microbial population contained within the composter is fairly stable, and made up mainly of bystander bacteria and fungi. They eat happily, and most of their by-products, while perhaps a little smelly, have no negative impact on our health. Exposure to these species may also help to increase the diversity of our own microbial population.

Once you add meat, fecal matter, or diseased plants, however, the microbial population changes dramatically and the risk to human health increases. The problem is exposure not just to foes but also to their toxic by-products. The number of hostile microbes will grow to high levels if the nutrient supply is sufficiently rich, creating a threat not unlike that of a microbiology laboratory, where gloves, masks, and lab glasses are required. Even the tiniest composting containers can be a problem. Better to dispose of these toxic items in the regular trash or in municipally provided organic waste containers.

Of course, municipal disposal means these items will eventually end up in a different type of composter, and this too can be an issue. Municipal biocomposting facilities are excellent at reducing landfill, but they come with their own risks for health. The sheer size alone means an even higher concentration of microbes, many of which will produce toxins and other harmful chemicals. If the facility is aerated, there is a high potential for these troublesome molecules to spread on the wind. This could mean trouble for populated areas downwind, particularly in people with chronic conditions such as asthma and skin irritation.

That's why I hope municipalities are taking measures to ensure that toxins are not given access to the human population.

Those measures could include installing filters on site and placing biocomposting facilities well away from urban areas.

THE NEXT PAGE A bacterium called *Gluconacetobacter xylinus* may one day help to conserve the world's trees. When put into a culture with nutrients such as glucose, this bacterium creates thick mats of cellulose. When dried, these mats become as thin as vellum. The sheets are fully flexible and can be used for writing or printing. Imagine—all the paper we need, and not one tree cut down to make it.

And for writing on microbial paper, what better than microbial ink? There are quite a few species capable of making some of the most delightfully coloured pigments. Want a nice yellow? *Pseudomonas* can do that. How about orange? *Brevibacterium* will suit your needs. As for the standard blue, look no further than *Streptomyces*. There's even a microbe capable of producing that editorial red. It's an algae called *Dunaliella salina*.

There's another advantage to going microbial for your next writing assignment. If you're not happy with what you've written, you don't have to throw it in the wastepaper basket, you can pop it in your mouth. Both the inks and the paper are perfectly safe to eat and offer an excellent source of fibre. It brings a new meaning to the expression "read and inwardly digest."

Of course, if you go looking for the microbial aisle in your local stationery store, you'll be disappointed. These items are still years away from mass production. Yet considering the environmentally friendly means by which they are made, they should become very popular. Not only will we be able to express ourselves as we have for centuries, but we will be able to do so without worrying about tree loss, chemical pollution, and landfill buildup.

MAKING SENSE OF GMOS People on both sides in the GMO debate—for and against—tend to be adamant and uncompromising. But genetically modified organisms are not all good or all bad. Some are excellent for our health and help us combat serious diseases. Others appear to be nothing more than line items in a corporate strategy. Figuring out the difference is not easy, but an all-or-nothing approach is not particularly helpful.

The practice of genetic modification isn't new, biologically speaking. We are all genetically modified organisms, in that we are not direct clones of our mothers. The differences in our genome give us unique traits; they're what make us individuals. With the discovery of DNA in the 1950s, we have been able to determine the nature of those differences and how we can harness them in the lab.

Today, we can effectively mimic naturally occurring genetic modification. The only difference between natural and manufactured modification is how long each takes. Certain processes, such as resistance to chemicals, improved lifespan, and increased nutrient production, may take years to millennia to develop naturally, depending on the species. Using laboratory techniques, we can make the future the present.

But we can do more than just speed up Mother Nature's timeline. We now know how to insert genetic elements into organisms to give them completely different traits. This isn't some sinister science fiction story being turned to fact. We've simply taken hold of natural processes.

Consider this: mammals are by definition genetically modified. As I explain later in this book, our branch of the biological tree came to be only after a virus was inserted into our genome. Without this natural form of genetic modification, we simply wouldn't be here.

Genetic modifications are also happening regularly in the wild. It's a common practice for viruses and bacteria to exchange genetic material, for example. This can actually help the recipient stay alive in an otherwise unwelcome environment.

So what is the difference between Mother Nature's genetic modification and that of a microbiologist? It's the control factor. In the lab, we have full command of the exchange and can ensure that we produce exactly what we want.

The laboratory practice of genetic insertion is also known as genetic engineering. It's been a welcome advancement in medicine. Microbial GMOs have proven invaluable in many ways—increasing, for example, the availability of life-saving drugs such as insulin and blood-clotting factors. These medicines once had to be taken from the recently departed, meaning their supply was limited. Today, they are made in abundance using bacteria and yeast.

Genetic engineering has also been the basis for a number of highly touted vaccines and therapeutics for some of the worst pathogens, such as Ebola. Though you may have heard of them in the media, you might not know they were created and are now produced by genetically-modified microbes.

But that's not all. Thanks to this practice, we will one day be able to create personalized medicines. Necessary therapeutics will be identified for a single person and then, thanks to engineering, designed with the individual in mind.

Though the use of GMOs in medicine has been invaluable, in agriculture, genetic insertion has been met with mixed reviews. There are humanitarian reasons for creating medicinal GMOs, but the modification of plants is not always philanthropic. While some food choices, such as rice, have been altered to improve their nutritional value, others have been designed to resist pesticides and other unnatural chemicals. This particular practice has few actual benefits for us.

What makes the situation even worse is the lack of control over the spread of these inserted elements. In the lab, cross-contamination is controlled quite easily. In the wild, it's almost impossible to do, particularly in crops that cross-pollinate. This can lead to the resistance being passed on to other plant types. How rapidly and effectively this occurs is still relatively unknown. Even more uncertain are the current options for dealing with this phenomenon. In essence, no one really knows how to control a GMO.

This is perhaps the crux of the GMO debate. It reminds me of the story of Victor Frankenstein. He had the power to advance medicine by bringing the dead back to life. But he had no idea how to stay in control of it. Once his innovation broke free of the lab and went into the wild, chaos ensued and the townspeople reached for their pitchforks. When confined to the lab or limited to a certain species, GMOs are fine but those that spread out in the wild are for the most part monsters.

It's clear, then, that the actual issue with GMOs isn't modification—that happens all the time in nature—but controlled modification, or engineering. To alleviate people's fears, this type of scientific control has to be itself controlled. But stopping genetic modification altogether is not an option, as it would prevent valuable medicines from being developed and health breakthroughs from occurring. Instead, the practice needs to be regulated so it is done responsibly.

In order words, all genetic-engineering practices should meet three criteria: they must directly improve human or environmental health; they must be reproducible and available to other researchers; and they must not cause cross-contamination in the wild. This may seem like a tall order, considering some of the controversial GMOs already in the market. But moving forward, heeding these three tenets can ensure that we improve some of our greatest health deficits while we avoid doing lasting harm to the environment or ourselves.

OIL CHANGE No matter what you think of the taste, fish oil is good for you. It contains omega-3 fatty acids, which have incredible health benefits. The oil is also good for livestock, as it helps them stay healthy and maintain good weight.

But there is a major problem: the availability of these fatty acids is dependent on fish stocks, which are dwindling. Farming has helped to maintain the number of fish available in the market, but this hasn't helped the oil situation. That's because fish, like humans, cannot manufacture omega-3 fatty acids. They have to get it from somewhere else—namely, algae. In the wild, these microbial plants are abundant in low concentrations. In fish farms, however, they can overgrow and quickly become a contaminant. Fortunately, in 2015, researchers in the UK figured out an inventive method to increase the yield of omega-3s. Instead of relying on algae, they decided to take algal genes responsible for producing the fatty acids and put them into a weed known as false flax.

There were several reasons behind the choice. False flax already made small amounts of these vital nutrients. The only difference between the natural plant and the genetically modified version would be higher levels of the fatty acids. The plant also didn't cross-pollinate, so no other crop would inadvertently end up with the modified genes. Finally, it could be grown almost anywhere on the planet, meaning it could be used to support local populations.

The process of making the new flax was for the most part basic engineering. The genetic material responsible for making fatty acids was taken out of algae and put into the flax. Once the insertion was complete, the concentration of omega-3 fatty acids in the flax went from a few percent to almost twenty. In some crops, the yield was so high the oil could be used to feed both livestock and farmed fish.

The potential benefits of this innovation are enormous. And it also meets the three criteria of responsibility I mentioned on page 46. First, the newly formed plant has a direct positive impact on animal and fish health by providing a vital nutrient. (It also helps to reduce the squeeze on fish-oil stocks from the wild.) Second, the engineering technique is simple enough such that the process can be duplicated by any other researcher capable of performing the same protocols. Because the plant can be grown anywhere in the world, there's no monopoly on the end product. Third, there is no chance for cross-contamination. This way, no other crops will mysteriously start producing high levels of omega-3 fatty acids.

THE CURSE OF CLIMATE CHANGE The earth is undergoing a transformation. Temperatures and humidity levels have been slowly increasing.

And with warmer and wetter temperatures, many microbial species are not just thriving but expanding into parts of the world they once found inhospitable. Some of these microbes are pathogenic and very bad news for humans. Take, for example, *Vibrio parahaemolyticus*, a bacterium named for the way it uses vibration to move and for its ability to break down blood. When humans ingest the bacterium by consuming raw shellfish, they get sick with diarrhea that sometimes can be bloody. The bacterium can also get into wounds and cause life-threatening septicemia. *V. parahaemolyticus* can only be found in warm waters, as it needs a temperature of 15 degrees Celsius to survive. This previously meant that only seafood found in tropical and at times temperate zones were at risk of carrying it. But over the years, as the temperature of the oceans has risen, the bacterium's geographical range has widened. It is almost at pandemic levels in various parts of the Far East and has been found as far north as the waters off Alaska. Thanks to climate change, the bacterium has moved to new homes and presents a very serious risk.

It is almost impossible to rid the oceans of this bacterium, meaning our only options are to cook our seafood or avoid potentially contaminated raw seafood altogether. Cooking most types of shellfish should not present a problem. But seafood such as oysters and shrimp, which at times are eaten raw, can pose a threat to health.

Microbial movement in the lab is usually measured in millimetres. This is a far cry from the hundreds or thousands of kilometres between planetary ecological environments. But these tiny organisms can still travel great distances by hitching a ride with a host as it migrates. When a larger organism moves from one area to another, it takes the microbes with it. This is quite common in the human world with the spread of diseases ranging from tuberculosis to HIV. But human migration plays only a minor role in the widening presence of microbial pathogens. To understand the real impact, consider disease-carrying insects and and how climate change has increased their range, leading to the worldwide spread of some nasty microbes.

The most relevant threat to global health is the pesky mosquito. The bug's bite alone is annoying enough, but the real danger lies in the blood sharing, which can lead to diseases such as malaria, dengue fever, West Nile virus and Chikungunya virus. At one time, the mosquitoes responsible for passing on these diseases were not able to venture into North America and Europe; these places were too cold, particularly in the winter. But now mosquitoes are working their way north. The movement has been slow but significant, and many once-safe regions could soon could soon become hotbeds of mosquito-borne illnesses.

The tick is an even bigger threat. This tiny insect harbours a bacterium called *Borrelia burgdorferi*, which causes Lyme disease. This is an enigmatic condition with a variety of symptoms. In the short term, the bacterium can cause skin rashes, fever, chills, and fatigue. But if the infection is not treated successfully with antibiotics, a person can suffer years of complications ranging from joint and nerve pain to cardiac disease and brain dysfunctions. Some sufferers end up with paralysis.

Twenty years ago, this infection was rare because ticks require a fairly specific climate of high humidity and temperatures above 7 degrees Celsius. But the combination of warmth and increased water in the air has given them the ability to wander and infect. In a generation, Lyme disease has become increasingly common and is now a priority for a number of countries once thought to be safe thanks to their cold winters.

These three examples are just the tip of the melting iceberg when it comes to the impact of climate change on microbial health. They're the ones that have taken up most of the headlines over the years, but many other diseases are also migrating and causing troubles. The risks to our health are increasing, and even more dangers are on the way. As we simply cannot halt the spread of microbes, we need to learn to deal with the risks and develop new defences such as vaccines to keep us safe in a more threatening world.

Parainfluenza virus

3. HYGIENE

VARIETIES OF SWEAT Body odour is caused by a combination of sweat and bacteria. Sweat keeps our skin moist and regulates our temperature, but it also contains fats, hormones, and proteins to strengthen the barrier provided by our skin. These molecules are also used as food by the bacteria there. When they are done eating, these species release waste products—their own fecal matter—right onto the skin. Many of these have an aroma; some are sweet and musky, but most are quite unpleasant. The determining factor isn't the bugs, however, it's the type of sweat they are given. Humans produce four varieties of sweat, each one serving a different purpose. Depending on their content, the potential for smell can range from minimal to downright awful.

The least troublesome types of perspiration are thermal and gustatory sweat. They are produced when our bodies get too warm from the weather or exertion. The latter, more commonly known as the "meat sweats," is caused by increased blood flow to the stomach when we eat dense food. Unlike thermal sweat, this type of sweating is limited to the forehead. These varieties are not associated with significant smell, since they consist mainly of water, electrolytes, and small quantities of waste products, primarily urea and ammonia.

The other two forms of sweat, emotional and apocrine, are the ones that cause all the trouble. It's because these provide social cues to others about our health and welfare, both biological and psychological. Emotional sweat usually occurs on the hands and in the armpits, the genital regions, and the feet. Apocrine sweat is similar to scent secretions from animals and is designed to keep enemies at a distance while inviting potential mates to get closer. (The word apocrine comes from the Greek *krinein,* "to separate.")

Emotional and apocrine sweats are thicker than thermal sweat because they are rich in molecules such as fats and hormones, many of which have a mildly pleasant aroma. Some of these secretions will be used by the skin to maintain elasticity and bonding. But most will be taken up by bacteria, used as nutrients, and eventually shed as smelly by-products. Over time the strength of the smell will grow as the concentration of the bacteria and the by-products rises. Eventually, without intervention—washing—the odour will emanate and find those sensitive olfactory nerves.

How often you need to clean comes down to the relative concentrations of the four types of sweat. Thermal and gustatory sweat will have little impact on your scent even if the sweating becomes profuse. The lack of any significant nutrients will keep the microbes from causing any olfactory offences.

But even the slightest amount of apocrine or emotional sweat can eventually turn a person's scent foul. The stench is essentially a call for help from a suffering body, and it signals to other people that an individual is experiencing abnormalities such as infection, fear, and even depression.

As to how often you can go between washes, ask someone you trust. Let that person get within a foot of you, inhale, and then decide. Whatever you do, don't trust yourself. Our olfactory systems have a tendency to become desensitized to our own smells and will be unable to pick up on any offending scents until they are overbearing. Essentially, if you can smell yourself, it's already too late.

HOW TO AVOID LAUNDRY Bacteria will inevitably get into the fabric of the clothes we wear. If the environment is dry and cool, the bugs will most likely end up dying (or at least not growing). But if the humidity level rises above 60 percent and the temperature creeps towards the 30- or 35-degree Celsius mark, many species will begin to eat whatever food happens to be around—such as dried sweat and skin fats—and multiply en masse. Give the tiny creatures as little as a few days and they can foul a perfectly good piece of clothing with an unappealing odour.

Even if a piece of clothing is rancid, though, there is a way to save on the cost of laundry and make it clean again. The trick is to change the temperature from one that encourages growth to one that stops it altogether—and may even kill most of the residents. This can be accomplished by using hot water.

But there is a problem. A lot of fabrics—and most shoes—are ruined by washing in hot water: polyester and silk; leather, of course; and raw denim jeans, which actually come with instructions from the manufacturer saying they'll end up looking their best if you don't launder them at all for the first year or so of wear.

So if hot water is out, what can you do? Well, the best way to remove microbes and their by-product smells from these fabrics—as the raw denim community knows well—is to put them in the freezer. (Works for shoes too!) Most microbes can survive cold weather, turning dormant until the environment becomes more hospitable. But once the temperature drops below –10 Celsius—the standard temperature for most residential freezers—the bacteria have a difficult time staying alive. When ice crystals form inside the cell, they can rupture membranes, destroy critical pieces of cellular matter, and disrupt the genetic material. Give your clothes a good freeze for a few days and you'll reduce the microbial load significantly and also the smell.

Larger items like blankets, duvets, and pillows can be bundled into garbage bags and left outside—preferably nowhere near the actual garbage—during the winter months. Two days in frigid temperatures will leave them free of not only microbes but also other pests such as mites.

THE ARMPIT OF OUR EMOTIONS There are two easy ways to find out if someone is stressed. The first is to ask. The other is to sniff that person's armpit.

The armpit is scientifically known as the axillary region and has been the focus of many a study on human psychological states. The decades of research have revealed that a trained nose can detect anxiety, fear, calmness, and even attraction.

Only a small percentage of the malodorous chemicals in the armpit are actually made by our bodies. The rest are made by the microbes that live there. This warm, moist area is a haven for bacteria, and they enjoy living, eating, and, well, defecating there. It's that latter part that causes the smell.

Depending on the person, a wide variety of bacteria may call the armpit home. Some of these produce sickly-sweet aromas; others make woodier and muskier scents. Some release odours that smell like cheese. Then there are those that offer the earthy odour of a damp forest. The determining factor is what we feed them.

Depending on our emotional state, we may form different types of fats and hormones in different concentrations. The best example of this comes when we are stressed. The main biological signal of stress is adrenaline, and it sparks the entire body to go on the defensive. Being on the defensive produces higher amounts of sweat, a higher body temperature, and a greater number of chemicals (which will give off the smell of fear once broken down by the resident microbes). Within minutes to hours, our scent can change from pleasant to offensive. It may take days before the system calms down and we return to a more natural, pleasing aroma.

DEODORANT VS. ANTIPERSPIRANT Which kind of underarm protection is best? Deodorants are specifically designed to target the bacteria on the skin. They don't necessarily kill, but they can inhibit growth so these microbes cannot produce an overwhelming odour. Most also have some form of fragrance to mask any bacterial unpleasantness that may develop over time. Deodorants are quite efficient, although they don't tend to last very long. They also don't prevent sweating, so those who have an overactive sweat gland could still end up with wet marks on their shirts.

The chemicals in antiperspirants differ depending on the brand, but almost all contain metal aluminum. This element gets into the skin and blocks pores to prevent sweat from escaping. Instead, excess water is sent to other regions of the body for excretion. This is a highly effective way of keeping your armpits dry. There may be hazards, though. Some researchers have suggested aluminum in the blood increases the risk for some cancers and Alzheimer's disease. But no concrete links have so far been found, however, meaning this may simply be a matter of coincidence, not causation.

The other issue with antiperspirant is that it cannot fend off certain types of very smelly bacteria. When antiperspirant is used constantly, some environmental types of bacteria can find a home by attaching to the chemical components of the cosmetic. One such example is a group of bacteria found in dirt. We pick these hitchhikers up on our hands and no doubt transfer them to the armpits during our daily checks. Once they have made a home, they can use the sweat to thrive. For us, this means a change in our odour to include the smell of, well, mud. This is reversible by quitting antiperspirants for a while and going with either a deodorant or your natural scent.

OLD SPICE It happened to me soon after I turned forty. The person walking beside me noticed I had a particular smell.

I checked my armpits to be sure, but I couldn't detect anything. So I asked what the aroma was like. She said, "Old-man smell." For anyone who hasn't had the chance to experience this odour, it's a combination of stale air mixed with grass and grease. It's not pleasant.

The chemical behind this particular aroma is a compound known as 2-nonenal. It's essentially a fat molecule formed as a by-product of various cellular processes in humans and bacteria. The chemical is better known in the food world as a component of fermentation, to increase the grassy and oily odour of aged foods such as beer and wheat. The chemical is also made in larger amounts in people over forty years of age.

Unlike other human odours, which are mainly caused by bacteria, this one happens to be our own fault. It's formed under the skin and released as a waste. You might be able to reduce the levels of the stuff by reducing your consumption of alcohol and tobacco, which tend to force the body to send out fats in much higher concentrations. You can also reduce your intake of animal fats and butter. These are high in omega-7 fatty acids and they can make for a rather stinky output once metabolized.

Showering and bathing temporarily removes 2-nonenal. A longer-lasting fix is to flush it out of the system with a lot of sweating. Getting a good workout can help keep the concentration of the smelly scrounger low enough to be undetectable to most people. It may take a few days to get there if you've been for the most part sedentary. Increased showering over this time is a must! Keep exercising regularly and you will soon begin to smell youthful and look it too.

JUST SAY NO Dermatologists tell us we are damaging our skin with excessive use of soaps and detergents. The harsh chemicals they contain do slough off dead skin and take aromatic molecules with them. But we also lose quite a number of our native bacterial species. Some of the bacteria on the skin are needed to help keep odours at a minimum. They eat up ammonia and urea—chemicals that make us reek. But they also form a chemical called nitric oxide, which is one of the most important molecules for overall skin health.

Nitric oxide, or NO, is a very small molecule; it has only one nitrogen atom and a single oxygen atom. But its purpose far outweighs its simple structure. When we feel cold, NO opens up all the blood vessels so more warm blood can flow. NO is also an excellent antimicrobial, and can help fight off some fungal and yeast species. It can also help against acne, as it kills species commonly known to cause this skin ailment. Even when we have a cut or a scrape, NO will speed up the healing process and lower the chances for scarring.

Losing these NO-producing bacteria in the short term is not a problem. Most of them come back over time. But there are certain species that need to have a combination of sweat and dead skin to thrive. When they are continually starved due to washing, they end up disappearing. This can lead to an accumulation of ammonia and urea, both of which can give us an rather nasty odour.

The best way to increase the number of bacteria making NO is to sweat and let it stick around. Sweating brings ammonia and urea to these bacteria so they have the food they need. Unfortunately, this means sacrificing routine washing, which can lead to an increase in other smelly chemical by-products.

There may be a solution to this quandary. Instead of trying to let the natural bacteria on the skin grow, you can add it in the form of a spray or a bar of bacterial soap. The manufacturers of these products claim that users need to wash less often and may never have to use regular soap again.

That may be so, but nothing beats the feeling of lather, so it's unlikely soap will ever be replaced entirely. Just be sure to avoid antibacterial soap containing triclosan. These products are no more effective than regular soap and pollute the environment once they go down the drain. Natural soaps leave you fresh and clean, and with the bacteria you need to keep your skin soft and healthy still in place.

GOOD FOR THE SOLE Those yellowish-white areas on our feet denote the presence of accumulated dead skin. For some of us, that yellow tinge can cover the heels, the sole, and the toes. Without a pedicure, these areas can become unsightly and also pose a potential health hazard. The accumulation of dead skin offers both bacteria and fungi a wealth of nutrients, as well as protection from the outside world. These microbial foes can huddle inside the layers and grow without interference from the body's defences or the gentle scrubbing of daily washing. Although most bacteria will only generate smells, some can cause worse problems in the form of skin infections ranging from cellulitis to athlete's foot to a form of gangrene. When these infections take hold, no amount of scrubbing will help; you'll most likely need medical attention.

One of the best ways to prevent the accumulation of dead skin is to ensure your feet are properly cleaned. This means not only giving them a wash every day but also taking some time to scrub off those skin cells at least once every week. It can be done with a pumice stone, an exfoliation board, creams containing rough kernels like apricot seeds, and even specialized creams made to dissolve dead skin. You could even try an Asian remedy called a fish pedicure, in which fish nibble on your feet. But the best option might simply be to go back to an ancient remedy: dissolve some Epsom salts in warm water and soak the feet for about a half hour. Some of the dead skin will fall off and what remains will be easily removable. It's efficient, easy to perform, and can help prevent the nasty problems associated with unfriendly bacteria and fungi.

FRIENDS OF OUR FEET Foot odour comes in four main varieties: sweaty, cheesy, vinegary, and cabbage-y. That's because of chemicals produced by the bacteria down there. Methanethiol is a key component in the flavour of cheddar cheese. Acetic acid is a result of sugar fermentation—and is better known as vinegar. By-products associated with rot, such as propionic acid and butyric acid, can leave feet smelling like rancid cabbage. The most common foot-related chemical, isovaleric acid, is responsible for the smell we call "sweaty." Our noses are up to two thousand times more sensitive to this chemical than the others, and many of us can recognize it even at the slightest concentration.

Only a few types of bacteria have learned to enjoy inhabiting the foot. Most of these are friends, despite their smell, and our lifelong partners. At any given time, we have hundreds of millions of them living happily on our feet, which they regard as the perfect environment: warm, moist, and offering an unending supply of nutrients in the form of dead skin cells. They adhere to us shortly after birth and stay with us for the rest of our lives.

They are also a necessary part of keeping our feet healthy. The bacteria release oils that help keep skin soft and enzymes that break down dead skin and prevent dry, flaky areas, as well as calluses. Our foot friends also provide a barrier against microbial pathogens. Our bacteria are very territorial, and they have mechanisms to ward off disease-causing visitors. They produce a number of defensive molecules, called antimicrobial peptides, which seek out and kill any invaders. These molecules are similar to antibiotics, but pathogens cannot develop resistance to them. To have the healthiest feet, we need these good microbes working hard for us. It can be difficult to assess their presence with our eyes, but we can always perform a smell test to ascertain if our feet are in good microbial hands. When we have a smell that is familiar to us—even if it isn't pleasant—we can be sure we're maintaining the same microbial population.

If that smell changes, though, and becomes more bread-like, grape-like, or acrid instead of sour, it can be a warning sign. There are several infections, mainly fungal, which can take residence on the foot and start to attack. Unlike our microbial flora, which prefer to feed off dead skin cells, these intruders want to eat something fresh. Without proper treatment, these pathogens can cause rashes, breaks in the skin, and larger wounds. Should this happen, you may require medical attention.

While the smell of your feet is usually a sign of your overall health, it might not do wonders for your social life. Thankfully, there are ways to keep the friendly bacteria happy while still keeping scents to a minimum. One option is to use talcum powder or charcoal inner soles. They both absorb the smelly chemicals and prevent them from dispersing in the air. While they won't make your feet smell nice, they can keep your shoes from accumulating noisome chemicals.

There are other naturally derived compounds—including citral, geraniol, and limolene—that are known to help improve that familiar foot smell. These chemicals shift the way the bacteria make by-products, inhibiting isovaleric acid from being produced in the first place. They can be found in several common foot-care products available in drugstores.

SMELL YOU AGAIN SOON We've become accustomed to certain common practices for greeting each other—handshakes, hugs, kisses, and chest bumps (for those engaged in bromances). All of these are for the most part culturally accepted and also have little chance of offending with odours. Yet some cultures tend to regard the smelliest areas of the body as critical to a proper greeting.

In India, elders are greeted not at the hands but at the feet. This is a sign of utmost respect. In essence, you are saying the most dishonourable part of the person you're greeting has greater esteem than any part of your own body. The microbial component of this exchange is not really given any thought, but those well-travelled feet might actually expose you to a host of new germs.

In Australia, a custom of the indigenous Yidindji peoples involves the armpits. As a form of respectful greeting, an elder rubs his pits to soak up the aroma and then places his hands on the face of the visitor. The scent is said to give a blessing to ensure good travels.

If neither of these rituals appeal, you could try a third traditional gesture of greeting: nose pressing. It's performed by a variety of people all over the globe and consists of rubbing the nose all over the face of another person and inhaling the aromas. It's a perfect way to demonstrate intimacy between two people without taking it to a romantic level.

The more widely practised handshake is neutral, simple, and offers no aromatic cues—or so it would seem. But in 2015, a group of researchers from the Weizmann Institute of Science in Israel secretly filmed 271 people to investigate their post-handshake activities. They found that 55 percent smelled their hands afterwards. What's the reason behind this action? The researchers suggested that humans still want to smell their way into a relationship.

HAND OF FRIENDSHIP When I'm lecturing, I always stress that staying friends with our microbes is crucial for our well-being. If that's the case, someone will inevitably ask, why do I use hand sanitizers? After all, they kill the friendly bacteria and could end up doing more harm than good. It's a valid question. These products are indeed designed to kill bacteria, viruses, and fungi indiscriminately. Surely some of the good bugs will end up getting taken out with the bad. But there is a way to enjoy the best of both worlds.

The whole point of using a sanitizer is to keep the hands safe from transient microbes. We pick them up everywhere we go as we touch the world around us. They won't establish residence on the skin, though. They are only taking a free ride to a place where they can have a better chance to live. For pathogens and other infectious diseases, that optimal place is inside the lungs or the intestines. When you touch someone (or even yourself), you're always running the risk of giving a free ride to a pathogen. This happens far too often and causes up to 80 percent of the infections we get.

Handwashing is the best solution, but we aren't always within easy reach of a sink, running water, and soap. So hand sanitizers are the next best thing. A lotion, gel, or foam containing 62 to 70 percent alcohol—preferably ethanol—will remove most, if not all, of the various microbial species on the hands in twenty to thirty seconds. It's a rapid and efficient way to ensure that pathogens and other infections don't spread. In the healthcare and food industries, this is a necessity. But there is that unfortunate consequence for our friendly bacteria. They are as vulnerable to the lotions as our foes and will end up dying in the process.

This shouldn't be a concern, though. You can quickly regain the good bacteria on your hands. All you need to do is touch an area of the body where only the good microbes live. One of the best places is back of the upper arm, the triceps area. Give a couple of good rubs and those friendly residents will happily make the transfer. The skin is once again repopulated, and you can go on living without worry for your microbial friends until the next round of sanitization.

THE INCUBATOR UNDER YOUR NOSE In the laboratory, bacteria are grown using a standard technique. They're placed in a test tube or on a petri plate in a nutritious liquid known as a growth medium. Then the plate or tube is put into an incubator—a dark, warm environment with plenty of air—where within a few hours the original bacteria can produce millions more.

Humans also have an incubator, known as the mouth. This cavity has everything a bacterium needs to thrive. Overnight, the tens of millions of bacteria our mouths contained at bedtime can increase to hundreds of millions, if not billions. At any one time, we have between one hundred and three hundred different types of microbes residing everywhere in our mouths, from our teeth and gums to our tongue and cheeks. Most are good for us, but some can cause problems, including nasty breath.

The most common bacteria in the mouth are relatively few in number and belong to a group known as streptococci. Some are involved in the formation of gingivitis and cavities; others can cause strep throat. But the majority help us maintain a healthy mouth by forming protective layers called biofilms on our teeth and gums. These not only shield us from harmful bacteria but also help to prevent plaque buildup.

Other types of mouth bacteria give our breath a putrid smell. Though they are not pathogenic, they tend to use nutrients differently than streptococci. They also have different waste products, many of which have a distinct smell, for example rotten eggs (hydrogen sulphide), rotting meat (methanethiol), rotting fish (putrescine), feces (indole) and smelling salts (ammonia). Any one of these is bad enough, let alone a combination.

For many of us, bad breath can be swiftly eliminated with a toothbrush and toothpaste. Yet there are millions of people who simply cannot get bad breath to stay away. They suffer from a condition known as halitosis, which is caused by a catastrophic change in the mouth's microbe population. When these stinkers take up residence in the mouth they are hard to evict. Brushing won't help and nor will mouthwash.

So what does help? One treatment is a disinfecting chemical known as chlorhexidine. It is effective against not only these foul bacteria but also other long-term problems, such as plaque. Another is to use higher amounts of fluoride and zinc salts to remove these bacteria and encourage good species to grow back.

Then there are antibiotics, which are designed to kill off bacteria. They have drawbacks, however, as they do not discriminate between friends and foes. They sometimes take the good out of the picture along with the bad, leaving not only a lack of diversity but an opening for unwanted future residents. As food, water, and other objects are placed into the mouth, new species, including other foes, will have a better chance to find a home. The simple way to prevent this massive shift is to avoid using antibiotics unless there is a true need for them.

One of the best routes to fresher breath is to fight microbes with microbes. Taking in good streptococci in the form of a supplement or even a microbial gum can reduce halitosis in a matter of months. Streptococci are very territorial and like to protect their own, including our gums and teeth. If they have the numbers, they will chase off the odorous bacteria or kill them outright.

To keep our oral health and our social lives in the best shape possible, we have to stay attentive to our oral incubator and do our best to keep it filled with good bacteria. Apart from regular brushing and flossing and chewing the occasional stick of sugar-free gum, we should eat natural sugars such as fruit and drink at least six glasses of water a day.

CHEW ON THIS Chewing gum when a toothbrush isn't at hand is a common and effective practice. While the flavouring can help to keep bad smells at bay, the act of chewing itself can help to reduce the number of oral bacteria.

As soon as you put a piece of gum in your mouth, you begin to salivate. The increase in fluid helps to dilute the bacteria and send most of it to the gut. But the gum itself also has an incredible ability to bind up bacteria and hold them in place. In a matter of only two minutes, a hundred million bacteria can get trapped in the matrix. After that, the gum loses its ability to hold on to the bugs, allowing them back into the saliva.

What's most impressive, though, is not the number of bacteria caught up in gum's net but the types. Some good bacteria will find their way into the stretchy stuff, but most of the trapped bacteria are those we happen to pick up during our daily activities. Some of these can pose a threat to our oral health. Removing them can only improve the chances for a healthier mouth.

The best gums are those without any sugar or artificial sweeteners, and they are plentiful in stores all over the world.

It's pretty simple to obtain microbially friendly fresh breath: take a stick, chew away for two minutes, and discard. You're left with fresh breath, as well as a cleaner and healthier mouth.

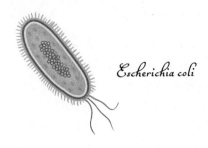

Escherichia coli

4. BEAUTY

MASCARA WITHOUT TEARS I admit it. I wear makeup for photo shoots and television appearances. And when it's time to sit in that chair I'm always a little concerned. Not about how I'm going to look, but about the risk of coming into contact with pathogens.

Bacteria love makeup because of its high fat content. In the container, these microbes can thrive, forming colonies and biofilms where their numbers can increase. Most of the species are harmless, but several potential pathogens have been found multiplying inside cosmetic containers. When the makeup is applied, skin and respiratory problems may result. The toxins produced as the microbes grow can cause rashes and itching, as well as nose and throat irritations.

But the worry is for the eyes. Bacteria tend to grow quite well in mascara (a type of makeup I *don't* wear), and over time they can become so concentrated they cause a range of problems, from simple irritation to significant infections.

To counter this issue, the cosmetics industry has introduced several preservatives and antimicrobial chemicals to prevent the growth of these harmful bacteria. While this does help extend the shelf life of the product, there are consequences. When applied to the skin, these antimicrobials kill all bacteria, good as well as bad, leaving behind a virtual wasteland. As we seen, that is no guarantee of safety.

Some companies have turned to microbes to resolve this problem, growing several friendly bacteria in the laboratory to form a liquid matrix known as the ferment. (Essentially the same process is used in making wine or beer, but the microbial version is non-alcoholic.) The solution can then be added to the cosmetic to provide a safeguard against unwanted microbial growth and lessen the chance of skin reactions.

Of course, the most effective way to prevent issues with cosmetics is simply to avoid using them regularly. But if wearing makeup is an essential part of your daily life, you should do your best to avoid having harmful bacteria end up on your skin or in your eyes. This includes replacing makeup items every six months, washing brushes after every use, and thoroughly cleansing the face nightly. The skin has its own ability to welcome the good bacteria and fight off the bad, but it's hindered when any of those makeup chemicals are around.

THE EYES HAVE THEM Microbiologically speaking, a contact lens is not unlike a petri plate in the lab. It's made from a similar matrix of synthetic chemicals and is for the most part incapable of allowing microbes to grow. But when a growth medium is added, the bacteria can thrive. In the lab, this nutrient base is comprised of a complex composition of nutrients and a solidifying agent called agarose. On the contact lens, the growth medium comes from the eye itself.

On the surface of the eye is a layer of mucus offering all the nutrition bacteria require to grow. But it's also home to a variety of immune defences that keep any unwanted microbial species from invading. When a contact lens is removed from the eye, however, the real-time surveillance for foes disappears as does the ability to fight them. This allows bacteria, amoebae, or fungi to grow using mucus as food. If unchecked, these microbes can form biofilms full of toxins and other harmful chemicals. Put that lot in the eye and the immune system will not be able to cope. Furthermore, contact lenses exposed to the environment can pick up travelling microbes. The air is filled with a variety of microbial species, and they will settle anywhere. Most of the time, they don't have the nutrients needed to grow, but when they arrive on a rich surface like a mucus-covered contact lens, they can quickly take up residence and, when put back on the eye, cause infection.

Putting contact lenses into a solution is the easiest way to prevent drying as well as infection. Salt water is best, preferably with a disinfectant added. The water disperses the buildup of mucus and other biological fluids and the disinfectant kills most microbial species. Before they are put back on the eye, the lenses should be rinsed with saline to remove any of the cleaning chemicals. The molecules responsible for keeping the lenses clean can also kill human cells, causing irritation and possibly damage to the cornea.

A BRUSH WITH DANGER Looking your best can be dangerous. We've already seen that makeup poses a threat—and so does getting your hair styled. The main suspect in this instance is the hairbrush. Without proper cleaning and disinfecting, it can be an agent of microbial spread and a source of infections.

Brushes are made to work on both hair and scalp to give your locks a smooth feel and appearance, but in the process they collect particles such as dead skin, dandruff, oils and grime, and of course, microbes. The amount of bacteria and fungi in the hair isn't all that high, but each time a brush is used, germs accumulate and can eventually pose a hazard.

There are numerous products available to keep your brushes clean. You can even make one at home by combining one glass of water, a teaspoon of dish soap, and a teaspoon of vinegar. Over the course of an overnight soak, the soap will break down all the dirt and grime, while the vinegar will kill off most of the bacteria.

Buildup is especially common if brushes are shared between people. If everyone using the brushes has a normal microbial population on skin and scalp, there's really no concern, but the brushes can also pick up bacteria from the environment, such as the washroom, leading to an accumulation of other species, including those found in feces. Bacteria love to feed on all that's left on a brush and can reach high enough numbers to cause a problem.

ATHLETE'S SCALP Scratching your head isn't as bad as some might want you to believe. It's a natural process of increasing blood flow to the scalp. But when the scratching is accompanied by the sloughing of white flakes, then there is a problem. What we call dandruff occurs when a fungus makes its home in the top layers of the scalp.

Dandruff is a specialized form of almost any skin condition caused by an invasion of fungi. On the toes, it's called athlete's foot; on the skin, the general term is dermatitis. From a microbiological perspective, the location of the infection doesn't really matter. The skin has gained an unfriendly new resident who is causing us harm.

These fungi are everywhere. You can find them at every corner of the globe, waiting around for some wonderful source of nutrients to come along. Animals and humans offer a rather plentiful supply in the form of sebum. This oily substance helps our skin remain moist and also keeps the roots of our hair from getting too dry. It's a fantastic lubricant as well, and it makes getting into and out of our clothes that much easier. For fungi, however, sebum is nothing more than food.

Once a fungal species has made it onto the skin, it starts to form a home by essentially taking out parts of the dermal layer to make room. It does this by producing a number of enzymes known to break down the cells on the skin. It doesn't realize that this will eventually get the body upset. And when the body's annoyed, it sends in the troops to halt the invasion.

Although the battle will be microscopic, you will know about it because of one very annoying symptom: you'll itch. It doesn't matter where the infection is taking place, when combat starts the brain wants to scratch. It's a natural mechanism for the body to let the brain know there is an area that's in trouble.

The problem, however, is how long that itch can last—for days, weeks, and even months as the fungi and the defence forces do battle. The mild sensation can turn into a maddening situation requiring medical attention not only for the skin but also for the mind.

Treatments for these conditions vary and some may be found on drugstore shelves. Most are simple antifungal chemicals known to take out the invading species. But some are much more potent and require a prescription. In either case, the goal is to eliminate the invader and restore the integrity of the skin. These treatments are not always effective, and worse, they are temporary. After the treatment is over, the fungus has the opportunity to reinfect.

But a long-term solution may be found with natural skin bacteria known to feast on the same foods as the fungi. When these good bacteria are present, they tend to stick on the outside of the body and take in the sebum as it comes. They may even be able to form communities around the pores and hair shafts to ensure that any visiting fungus realizes there is no vacancy.

There are already some species known to help prevent fungal colonization, while others have the ability to kill off the fungi altogether. But their effect has only been seen in the lab. More research will have to be done before any microbial antifungal products appear on the shelves.

In the meantime, the best way to prevent or cure any fungal issues is simply to stay as clean as possible and use soaps known to break down sebum.

FACE FUNGUS Microbes love beards and find them an appealing place for growth. Each hair offers a perfect combination of surface area, nutrient availability, and moisture. Bacteria that live in the beard are mostly the same as those found on the skin and the scalp. But those bristles also provide a welcome stopover for certain other microbes. The coarse and curly hairs are perfectly designed to catch any microbes in the air, on food, or in the water. Most of the visitors will be harmless, but the chance of a hidden infectious agent is always present. Men with beards do tend to suffer more skin infections underneath, presumably because of poor hygiene and the overpopulation of certain species. These infections are usually minor and not contagious. As far as anyone has been able to determine, no one has ever spread a significant pathogen by way of a beard.

The situation does change, however, when beards enter risky environments, such as restaurant kitchens or healthcare facilities. In a setting where so many people's health is at stake—and indeed where infection can lead quickly to litigation—a beard could spell disaster. A beard net, better known as a snood, is mandatory.

If you have a beard, the best way to keep the risks low is to wash it regularly, tug it and pull it, and massage the skin underneath, which helps to rejuvenate the surfaces of the hairs and bring to attention any infections at the skin level. If you're living with a bearded soul, you need to keep your distance only if he has a respiratory illness. As with any other virally-infected person, it's best to stay at least two metres away until the symptoms have gone.

TAT ATTACK At its most basic level, a tattoo is a surgical procedure and should be done under strict hygienic conditions. When it isn't, it can result in anything from simple irritation to a rash requiring medical attention.

Once a bacterium is introduced into the skin, it may reside there inertly, causing no harm. But it may also find the warm, moist, nutritious environment too appealing and start to invade. The immune system will then defend the wound as best it can. Most species will die, but some have the ability to resist human defences. This only worsens the situation, as the immunological troops will fight even harder to remove the threat. These troops tend to huddle in a staging area, forming a nodule under the skin. Over time, this growth can deform the skin (and the tattoo with it). If left unchecked, these nodules can even cause rashes and lesions as the immune forces begin to attack the human cells in the fight against bacteria.

Bacteria are not the only potential invaders. Viruses can also find their way under the skin—and they can prove to be even more difficult to remove. The most common of these are the wart viruses. When they have access to human cells under the skin, they like to infect and hide. Unlike other more common viruses causing disease, those causing warts bring no major symptoms. Only months later, when the skin bumps appear, is there any real sign of the issue. Warts can be treated, but this too can lead to deformations of the skin and of the tattoo.

The other possible problem is fungus, as certain species have the ability to get under the skin and initiate an infection. Although this type of infection is rare, the complications can be problematic. The first sign of trouble is the formation of a bump on the skin similar to a pimple. At this early stage, it may be treated effectively. But if left alone, the fungus can grow in density, turn the bump into a cyst, and then eventually burst. This can leave a hole in the skin, and the tattoo may be ruined.

This can all happen when the immune system is healthy. For those whose immunity is compromised by HIV infection, chemotherapy, or even anti-inflammatory medication, the invaders can make things much worse. They attack the skin and the blood, which can be life-threatening if the pathogen spreads to the lungs, the gastrointestinal tract, or other organs. This is very rare, however, and can easily be remedied with proper medical attention.

Even after a tattoo has settled and all risk of infection is gone, the dyes may be a source of problems. Tattoo dyes, known as azo compounds, are safe for humans in low doses and are found in a number of products, including clothing, cosmetics, and food colouring. But bacteria tend to view azo compounds as toxins and will try to destroy them, breaking them down into harmless—and colourless—by-products. This isn't just bad for the tattoo but also for the person. Although these waste products are not toxic to human cells, they can cause another problem when they are absorbed into the skin: cancer.

Many by-products of microbial dye breakdown have the ability to enter a human cell and mess with its genetic material. When this happens, the cell could either die off, which usually happens, or become cancerous. In the latter case, there is usually nothing to fear as the immune system has a way of finding individual cancer cells and killing them. But when other risk factors are present, such as increased UV exposure or a weakened immune system, detection can be evaded and the cells can grow to a point where they can withstand the tumour killers.

WHAT THE BODY THINKS OF PIERCING The skin is our best barrier to microbial invasion, and any significant breach requires medical attention. But we have a tendency to pierce this bacterially impenetrable layer on purpose, for the sake of fashion. Normally, when we suffer a wound, our bodies respond in a very coordinated manner. First, broken blood vessels are clotted to stop bleeding. Then immune cells head into the now vulnerable area. When all seems safe, the cells responsible for healing start to rebuild the barrier. Depending on the depth of the cut, the healing process can take anywhere from a few days to a few months.

When the skin is pierced, the healing process is stretched to the maximum. This is no mere scratch or incision—it's a rather serious wound. Whether you want to pierce the ear, the nose, the tongue, the bellybutton, or some more sensitive region of the body, you must permanently remove a piece of the body to create the hole.

The skin cells now have to figure out how to heal around the ring or stud. This will involve redevelopment of the matrix. At the same time, the immune system must deal with a permanent invader, as well as the invitation to microbial infection offered by a long-term wound. At first, the entire area will suffer from inflammation. It's the best response in a dire situation. But as the skin figures out how to heal, the strong reaction will ebb and eventually disappear. The skin will once again regain its barrier status and all will return to normal.

For the first few weeks after a piercing, wounded area should be kept clean and devoid of any microbes, friend or foe. This can be done with antiseptics or alcohol, as long as it's between 62 and 70 percent in concentration. The area will have to be cleaned every few hours at first, and then daily as the healing process continues. The body will eventually figure out how to deal with the change and heal.

If there is any sign of an infection, medical attention should be sought. Until the skin has completely grown over the wounded area, bacteria can gain access to the blood. Should that happen, a life-threatening illness—sepsis—can occur. It's rare, but the consequences are too great to take any risks.

THE SAFE WAY TO TAN I enjoy tanning, both naturally in the sun and also in the salon. But I do it in moderation and try my best not to get too much ultraviolet radiation.

Our skin is full of what are called Langerhans cells, named after the nineteenth-century German scientist who discovered them. They work hard to keep us safe by identifying and targeting cancer cells. Unfortunately, Langerhans cells are very sensitive to UV light and exposure reduces their ability to hunt down tumours in the making.

For those who get only occasional exposure to UV light, the loss of Langerhans is not a concern. But if a person is a real sun lover or a "fake and bake" regular, then problems could occur.

Instead of just turning off their surveillance, the Langerhans cells will seek another place in the body to dwell. It doesn't take much to convince them to bolt either. As little as thirty minutes a day for two weeks straight can lead to their departure. It's not permanent and the cells will make a comeback, but slowly. It can take weeks for the body to return to normal. For anyone attempting a perpetual bronze glow, the UV exposure might force the cells to disappear from the skin for good.

This problem can be even worse if you choose to use a tanning bed. If they are not properly cleaned, tanning beds are a breeding ground for bacteria left behind by other clients. If any are pathogenic, they could lead to an infection. Normally, this wouldn't be a problem—the Langerhans cells repel the invaders. But when the Langerhans are depleted, it's essentially a disaster waiting to happen.

There's an easy way to stay safe: don't tan. But that may seem a little extreme. If you choose to get some colour, make sure to wait a few days between tanning sessions. This might be easy in a salon, but not so much outdoors. The best option there is to use sunscreen to keep UV rays from penetrating. Also, if you choose to take the fake way to being bronzed, try to use a standup area instead of a bed so you prevent any contact with a previously touched surface. If that's not possible, make sure to give the entire surface a good wipe with a disinfectant before you lie down. The bacteria will end up dead and you'll enjoy a tan without the worry of blemishes or infections.

SPOT OF BOTHER The skin has a hard job and pollution only makes it more difficult. Ozone, which is a primary factor in smog, is toxic to cells and can harm the outer layer of the skin. Car exhaust can accumulate, increasing the chances of skin cancer. Going indoors may not be much help; dust often has contaminants capable of causing redness and itchiness.

To combat the effects of pollution, the body produces large quantities of a chemical known as squalene, an oily substance providing extra protection for both the skin itself and the pores. For bacteria, it's a fantastic source of energy. When our bacterial population is diverse, the various species equally use up the squalene, allowing them all to survive and maintain their numbers. But when certain bacteria die off, others have the opportunity to overgrow in the pores or under the skin. With the right food source, they can form colonies and eventually be seen as bumps, better known as acne.

Acne is traditionally associated with teenagers, but thanks to pollution, people of any age can fall victim. The best way to prevent acne is to clean the face regularly to keep pollutants and squalene to a minimum. If you want to get rid of a spot, look for an acne product containing ingredients such as the antimicrobial benzoyl peroxide as well as an anti-inflammatory such as salicylic acid. These will improve the skin over a few days and return it to a smooth state.

ACNES VS. ACNE Acne is at its core an infection of the skin. Somehow, a bacterial foe has managed to get under the skin and make a home there. Its numbers increase—sometimes at a rate of millions per day—until the colony becomes a visible bump. The extent of the trouble really depends on the species. Some bacteria will cause little more than a small round mound that just asks to be popped. Others will produce cysts, which can't simply be pushed out as they tend to get deeper into the skin. Some can be removed only by a surgeon.

The troublesome bacteria are usually attracted by an over-abundance of a particular chemical on the skin. It's called a triglyceride, which means it's a mix of a sugar called glycerol and three fat types. We produce these molecules all over our bodies and release them through our skin as waste. When we wash with soap and water, the concentration goes down. But in some cases, pockets of this chemical will remain, particu-larly in the pores.

That's really where the trouble starts. Many bacterial species love triglycerides and use them as food. For them, a filled pore is a smorgasbord of goodness. As they gorge on it, they can spread underneath the skin, where they are protected against any disruptions ranging from touching to washing. It takes only a few days for the invaders to grow to levels high enough to be seen, and they seem to work best at night while we sleep. Once they are present, the only way to control them is through antimicrobial medication.

If the medication doesn't control the growth or the bac-teria head too deep into the skin, the immune system will get involved. This may reduce the number of bacteria, but the damage caused in battle can leave the area red, injured, and even scarred if the outer layer of the skin breaks. This can be helped in part by medications containing anti-inflammatories like salicylic acid. It's usually added into topical acne creams to help lessen the damage.

There are much more severe forms of acne. In these cases, the bacteria have not only gained access and grown, but also taken over the area, causing a war. Remedies are limited to antibiotics or hormonal treatments, which are not always effective. Each carries the risk of unfortunate side effects too.

There may soon be another way to prevent and treat these blemishes: bacteria. Some species are known to fight off acne and can help to prevent future outbreaks. And some of these acne-fighters are actually the same bacteria as those causing the troubles in the first place.

Here's how it works. Within a bacterial species there are several strains. Take, for example, *Propionibacterium acnes*. Its not entirely accurate name means "eats fat and causes acne." Some strains do just that. But others don't: they eat the triglycerides just like their harmful relatives but without producing chemicals that irritate the skin. They also don't annoy the immune system, meaning no inflammation, bumps, or zits. Instead, they form a balance with our bodies to keep skin healthy and smooth.

Researchers have yet to indentify the cause of the difference between strains. When they do, an acne-prevention medicine will be within reach—a living one containing good bacteria able to overwhelm the bad.

In the meantime, there are some easy ways to keep the skin clear. The first is to clean it by washing with a good soap to remove any excess triglycerides. The second is to regulate the acidity of the skin—this is where creams can help. It may sound strange, but a more acidic skin leads to less acne.

It's because the bacteria behind the bumps don't grow well in an acidic environment. It's one of the reasons why we see so many pH-balanced facial washes and creams. They essentially help to keep the face clean as well as free from unsightly bumps.

MICROBIAL COUTURE Many textiles are made from a single molecule, cellulose. It's a sugar found in all plants. Some fabrics, such as cotton or flax, come directly from plant material, but others, like rayon, can be made from extracted cellulose. The manufacture of these textiles is rather involved and requires significant amounts of plant material. As demand grows, more land is required to sustain supply. As the amount of available land decreases, the potential for a squeeze in the supply of these plants rises. As such, alternative means for textile production have been sought.

Some of the solutions have come from the lab. Spent textiles can be repurposed as clothing, paper, construction materials, and protective coatings on electronics. Making cellulose for the world of fashion is not so easy. But many species of bacteria have the ability to make cellulose and are coming to the rescue.

The concept of biologically grown textiles has been around for decades. All that's needed is sugar, glycerol, or some other organic chemical with the ability to be transformed into cellulose. Once these items are in place, you simply have to add the bacterium, yeast, or fungus of interest and wait. The process takes some time and a little optimization, but soon a film of textile-looking cellulose will appear. After drying, the fabric will be ready for use.

Mass production became a reality around 2010. Microbial clothing has started to appear—so far mostly in the form of works of art. But bacteria-based fashions may not be far behind.

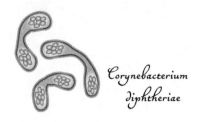

Corynebacterium
diphtheriae

5. ANTIBIOTICS AND PROBIOTICS

ANTIBIOTIC MASSACRE For decades, anyone dealing with an infection simply had to head on over to the doctor's office, get a prescription for antibiotics, pop a pill every few hours, and feel better. The process was quick and for the most part effective.

At the molecular level, an antibiotic recognizes and destroys molecules associated with a particular bacterial pathogen. In the lab, it's a simple process of adding the chemical and watching the bug die. But changing the landscape from the petri plate to the human body adds complexity and, unfortunately, risks. Getting the drug to hit its target is a matter of chance. It's why these drugs are designed to head straight to the bloodstream, where they can travel around the body hunting organs and tissues for enemy cells. This is incredibly stressful on the cells, however, and they go into a form of shock. This can cause an overall cellular dysfunction, which leaves the cell weakened and may even kill it, putting the integrity of a particular cellular environment at risk.

What antibiotics do to the body pales in comparison, though, to the massive effects they have on the microbial population. When these drugs head into the gut, it's a massacre. The number of microbes drops significantly, and many species are wiped out. The end result is a wasteland of dead microbes, with a tenacious few struggling for survival.

But that's not the whole picture. Not all microbial species will die. Some are used to dealing with antibiotics and have developed ways to keep themselves safe. They weather the storm and come out either unharmed or only mildly damaged. Once the antibiotics are gone, these species go back to business with one key difference: they have much more ground to call home.

The change in bacterial diversity is not a good thing, as many species known to survive antibiotics are our foes. Most aren't dangerous at the beginning, when they are happily being fed by way of their human host's diet. But as their numbers grow, they may start to attack the body in order to gain more nutrients. This is when trouble can begin. If we don't restore the diversity, the chances for infection grow by the day.

Once established, these bad bacteria can be extremely difficult to get rid of. Your options are to use more antibiotics (a vicious cycle) or to introduce diverse microbes from another source, such as another person's feces (This rather unappetizing therapy is known as fecal transplantation, and it's growing in use and popularity.) Another possible solution is to wait for diversity to be restored naturally. But this isn't a quick process. The gut may be left vulnerable for up to nine months before it even begins to return to a normal population. Even then, the balance may not be fully restored for another year and a half. In the meantime, any wandering microbe with the capability of overgrowth has an opportunity to come in and cause trouble.

There is a way to prevent the damage through the ingestion of friendly bacteria while taking antibiotics instead of afterwards. These species are found in various raw fermented foods and probiotics. When ingested, they help to provide some balance in the days during and after treatment. Just don't expect them to stick around, as they too are susceptible to an antibiotic attack. That's why they need to be taken every day at least two hours after one of the doses to keep their concentration high and consistent.

ANTIBIOTIC CRISIS It's incredible how our view of antibiotics has changed. They were the miracle drugs of the twentieth century, but today they are losing their lustre thanks to the rise of antibiotic resistance.

With each passing day, more pathogenic bacteria are gaining the ability to fight off these drugs. The process has been going on for decades, but now we are at a tipping point. Today, antibiotics are no longer the go-to for many infections. Many doctors now consider them a last resort. Unfortunately, infections continue to happen, and this squeeze on antibiotics has created a crisis. Without some type of change in the way we view and use these medications, we are doomed to return to a day when even the slightest infection could become life-threatening.

If you don't think this concerns you, think again. Resistance is inside you and me right now. Usually, this isn't a problem, as we are not continually taking antibiotics. But when we do, these already resistant microbes gain strength and eventually learn to tolerate the treatment entirely. We are essentially contributing to their resistance with each dose.

THE VICIOUS CIRCLE OF RESISTANCE

When we are prescribed an antibiotic, there's always the chance that the bacterial foes we're hoping to kill have somehow become resistant to the drug, leaving us in a rather bad position. We are still infected, but our arsenal to fight these pathogens has just grown smaller.

But that's not the worst of it. When a bacterium has figured out how to ward off an antibiotic, it can either keep that knowledge to itself or share the evasive procedure with its friends. If it chooses the latter option, everyone can enjoy the new-found defence. This process is called horizontal gene transfer. Essentially, the resistant bacterium makes several copies of its genetic material and then gives these copies away to other cells.

Gene-sharing among foes isn't limited to antibiotic resistance. Potentially harmful bacteria share a variety of genetic material to increase the viability of the entire population. This can help them survive in very unhappy conditions, particularly if the immune system is on the attack.

When the bugs feel threatened, they share defences so they can work together to weather the storm. For us, this does nothing but cause more trouble. Our immune systems are forced into longer campaigns that will eventually fail, and when that happens, we must seek out a doctor and possibly begin taking an assortment of powerful antibiotics. This perpetuates a vicious circle of increasing resistance. Until we stop using antibiotics or figure out how to prevent resistance—neither of which seems even remotely likely—we are going to continue to see even more failures of our once most trusted medical treatment.

ANTIBIOTICS OF THE ANCIENTS Long before the advent of medical schools, physician colleges, and formalized degrees, there were healers. They had their rituals, routines, and salves, many of which have been lost or forgotten. But some have stood the test of time, including the use of plants to treat infections.

Like humans, plants have to deal with a variety of microbes, both good and bad. But unlike their human counterparts, plants do not have a formalized immune defence system with which to protect themselves. Instead, they produce a number of antimicrobial agents, such as phytochemicals. Plant-friendly bacteria are able to resist these agents and hang around. But pathogens wishing not to reside but to invade find these chemicals toxic and end up dying in high concentrations.

Medicinal healers didn't know anything about molecular interactions, but they did know that certain plant extracts could be used to treat infections in humans. All that was needed was a willing patient, and there was usually no shortage of those. In a time when antibiotics were not available, unknown potions were the best hope. Most were useless but some were successful and have stood the test of time.

Now it seems there is a resurgence of interest in this form of traditional medicine. With antibiotic resistance rampant, alternatives need to be found. That's why researchers are combing through the books of antiquity to find plant-based recipes for testing in the lab. Not surprisingly, tea tree, oregano, clove, and cinnamon extracts have all proven their worth as antimicrobials in the lab.

Granted, at this stage, none have matched the effectiveness of antibiotics, and apart from garlic, which I discuss later in this book, they are not yet ready for medicinal prime time. But their promise is slowly turning into practice, and they are getting closer to being part of the next generation of antimicrobials. Eventually, some of these traditionally inspired concoctions will find their way back to our medicine huts, which today we call clinics and hospitals.

TECHNOLOGICAL CIRCLES Almost all cells on earth contain genetic material, the genome. But bacteria can have additional pieces called plasmids. These are not simply hitchhikers, though, as they can perform a variety of sometimes very helpful functions.

In nature, plasmids are usually used as weapons of war. They help the cell make toxins and other molecules known to harm other species. They also enable a bacterium to become resistant to antibiotics. They can even repair damage incurred in battle. Without plasmids, many species would be far less successful in their campaigns, and thus less likely to thrive in new environmental conditions.

There are hundreds, if not thousands, of different types of plasmids on earth, and the number keeps growing. They are so diverse because they mutate easily. They can gain and lose functions rapidly while still sticking around in a cell. For a bacterium, this offers the chance to test out different functions and keep only the ones found to be useful.

In the lab, the plasmid is one of the most important tools in genetic engineering. Different functions can be inserted into plasmids, such as the genetic code for the manufacture of a medicinal drug. This allows for a prototype to be produced and tested in a cost-effective way. If the drug is considered effective, the bacteria can then mass-produce the product for a fraction of the cost associated with making the chemical from scratch. If the genetic material isn't working up to expectations, it can be mutated further until the desired results are seen.

Most of today's plasmid-based medicines are limited to the lab. The bacteria produce the molecule of choice, such as an enzyme, or a hormone like insulin. Then it can be extracted, purified, and sold. This is where we are today, yet this is only scratching the surface. With proper engineering, plasmids could be designed to diagnose and possibly treat illnesses ranging from infection to inflammation.

The process is relatively straightforward. Once the plasmid has been engineered, it only needs to be inserted into a harmless bacterium capable of harbouring the genetic material. Then, through ingestion or another route, the bugs are given access to the human body and the product, whether a diagnostic chemical or an actual therapeutic is produced. The benefit is passively provided without any undue harm to the person.

Despite the relative simplicity of plasmid-based medical procedures, don't expect to see them offered as a treatment anytime soon. This type of medical advance may still be a decade or more away. Though the technology itself is simple scientifically speaking, the process for meeting global safety standards is not and requires years of clinical trials to ensure it can be used in humans.

INSTANT LONGEVITY? "How can I live longer?" It's an age-old question that I am asked regularly, and my answers usually disappoint. After all, longevity is the result of a combination of factors, including genetics, diet, lifestyle, and geography. All I can offer is an idea postulated more than a hundred years ago: keep friendly bacteria close.

That was the advice offered by Ilya Ilyich Mechnikov (he preferred to be called Élie), a microbiologist and immunologist who made significant finds in both fields. But even a Nobel Prize couldn't satisfy his ambition. He decided to become the enemy of death.

In the early 1900s, he recorded his thoughts in a book called *The Prolongation of Life*. It's a rather grim read as it focuses on aging and senility, the causes of natural death, and whether we should even think about extending life from social, moral, and philosophical perspectives. As he explored his observations, which he called optimistic studies, he mentioned more than a few times the influence of microbes. For the most part, he agreed with the commonly held view that microbes were associated with disease and death. The presence of one of these pathogens could hurt one's chances at living longer. Mechnikov was writing before the discovery of antibiotics, so there wasn't an easy medicinal option for treatment. Essentially, an infection was always life-threatening.

But in one section, he performs a complete about-turn, suggesting that some germs may be both bad and good.

It appeared to him there was a way to prevent disease through the ingestion of fermented foods containing the preservative lactic acid. He wasn't exactly sure how this chemical kept disease at bay, but it seemed to, and he encouraged others to continue his research.

The idea took time to gain momentum. After all, most researchers thought it was madness to view bacteria in a good light. And as antibiotics became more popular and offered an easy means to maintain health, the concept of using good germs to fight bad ones fell by the wayside. Only a few impassioned researchers maintained their drive to keep Mechnikov's ideas alive.

Eventually, they began to see the fruit of their labours. Fermented foods were indeed proven to be good for health. Taking it one step further, researchers identified the bacteria responsible and determined that most of them produced lactic acid. These species were viewed as potentially aiding that goal of living longer. In the 1950s, the group was given an apt name: probiotics.

POPULARITY AND PERILS In the 1980s, a massive shift happened in Western culture. People became all agog about improving their health. About the same time, the awareness of good bacteria was slowly growing. By the turn of the century, people were beginning to think that bacteria could be used to make us healthier.

This public interest, while still generally low, was enough to attract the attention of international health agencies such as the World Health Organization (WHO). They recognized the importance of beneficial bugs, yet they also knew that the term *probiotic* was vague and could be abused. To counter this, an expert group of microbiologists formed a think tank to set some parameters on the uses of probiotics. Back in 2001, the experts came up with a definition to describe exactly what probiotics are: "Live microorganisms that, when administered in adequate amounts, confer a health benefit on the host."

Okay, it's still a bit vague. But, considering Mechnikov's assertion in the early 1900s about good bacteria prolonging life, this seemed to be a good start.

The worriers have been proved right. Many large consumer-based companies are riding roughshod over this definition. Walk down the aisles of your local grocery store and you will see—next to all the low-fat and gluten-free products—label after label proclaiming some version of "Contains Probiotics!" But the word *probiotic* on a box or bottle certainly doesn't mean the contents are guaranteed to improve your health. In fact, some probiotic products may do more harm than good.

Choosing an effective probiotic is not as simple as looking at the label. The regulations differ widely across the world. The only answer, I'm afraid, is to do your research and become an honorary microbiologist.

There are criteria to determine a good probiotic. The first is to know the types of bacteria in the product. If you don't see Latin, the product probably isn't going to help. The second is to know exactly what benefits can be gained from taking in these microbes. The claims can include improving digestion, improving the diversity of the microbial population of the gut, helping heart health, or preventing *C. difficile* infection.

This information can usually be acquired directly from the company, either on the product website or by giving them a call (yes, sometimes a phone is the best means of gaining information). If no claims are being made, then don't expect any benefit. Finally, you have to know if you are getting enough. Research has proven that adults need at least ten billion bacteria in each serving to make any difference. For the record, yogurt only contains millions per serving.

This may seem like a great deal of effort, but it really isn't. If you cannot get all the information in a matter of a few clicks or a short conversation, simply move on. If a probiotic is really meant for your health, the knowledge you need should be at your fingertips.

THE PROBIOTIC FAMILY Of the millions of different microbial species in existence, only a few qualify as probiotics. Most of them come from one of two different groups, or genera. They are known as *Lactobacillus* and *Bifidobacterium*.

The original source of these genera is milk, or at least milk-fed babies (*Bifidobacterium* was first isolated from the feces of breastfed infants). Both genera are milk-loving and thrive when they are flooded with the various nutrients commonly found in milk, including lactose. They are the primary fermenters of yogurt and other solid forms of dairy. But they are not limited to this one food source. They have been found in almost every type of fermented product except wine—that's reserved for yeast—and have been isolated from areas all over the globe.

Within *Lactobacillus*, there are over two hundred different species, each of which possesses individual characteristics.

About a dozen qualify as probiotics. These are the most common:

- *Lactobacillus acidophilus*, which is known to create sour flavours thanks to its rapid production of lactic acid, a major chemical in fermentation and also one of the sources of umami flavour.
- *Lactobacillus rhamnosus*, which has a preference for another sugar found in milk, rhamnose.
- *Lactobacillus casei*, which uses many of the sugars and proteins found in cheese.
- *Lactobacillus plantarum*, which loves plants.
- *Lactobacillus reuteri*, which is named after its discoverer, Gerhard Reuter, and found in children during the first year of life.

Within *Bifidobacterium*, there are only fifty-one species, and only a few have been designated as probiotics. Many of their names are indicative of their carriers rather than their function; they include *animalis* (animals), *adolescentis* (teenagers), and *infantis* (babies). Others describe the way they look under the microscope, such as *bifidum* (having a cleft) and *breve* (shorter than the rest).

Lactobacillus and *Bifidobacterium* are the most tested forms of probiotics, and they have shown the most benefit to humans. Although each species has a unique function, they all follow a similar mechanism of action. When they arrive in the gut, they immediately go to work eating up and breaking down available food sources. This increases the efficiency of digestion and helps to minimize the time food stays in the intestines. As they chomp down, these bacteria release an assortment of nutrients, such as vitamins and minerals, which are then taken in by the intestines.

But helping with digestion is only half of the action. These two genera also make a variety of different chemicals, such as lactic acid. These chemicals are considered waste to them, but to us they offer a variety of benefits. Some help to maintain the integrity of the gut to prevent the onset of diarrhea. Others will interact with the immune system to maintain calm. Then there are chemicals with antimicrobial properties; they seek out and kill bacteria known to cause rotting and human infection.

Lactobacillus and *Bifidobacterium* may be the most commonly found probiotics, but they are not the only ones available. Other fermenters, such as *Bacillus*, *Enterococcus*, and *Streptococcus*, have proven their ability to keep the intestines happy and to promote balance in metabolism and the immune system. Many probiotic mixtures now include these bacteria.

When picking up a probiotic product, make sure to read the list of ingredients on the packaging. The list should be either exclusively microbial or at least have those species at the top of the list. If microbes are not the first ingredients—and something like sugar is—then the product probably won't be doing you much good.

THE HUMAN STRAINS Where do probiotics producers find their bacteria? The answer may be milk or some other raw material, which is then fermented. But sometimes, the bacteria come from human origin. This only ever means one thing: they come from feces.

It makes perfect sense. If you wanted a bacterium capable of doing good work in the gastrointestinal tract, where else would you look but the gastrointestinal tract? As to which humans should be sampled, there is no better choice than an infant.

Many bacterial species found in very young children are known to keep the gut in harmony and also to help the child grow stronger over time. Some of these bacteria also have the ability to keep pathogens at bay, making them even more beneficial. Researchers from around the world continue to explore the fecal matter of healthy babies in the hopes of finding the next rising probiotic star.

Identifying the bacteria best suited for a probiotic takes a lot of time and effort. Potential probiotics, regardless of origin, must go through extensive testing to determine if there is any good reason to include them in another person's diet. The most important criterion is the ability to use sugars commonly found in food. Because these bacteria are accustomed to only one source of food—breast milk, say—they may not be useful when exposed to other nutrients, such as fibre, fats, and meats. If they cannot rapidly digest these more varied foods, they won't make for a very useful probiotic.

The next criterion is the ability to survive the conditions experienced in the adult intestinal tract. Stomach acid is the first test; if the bacteria aren't able to survive here, there is no hope. Bile comes next. The small intestines are filled with these digestive chemicals and can readily kill bacteria not up to the challenge. The transition from the intestines to the colon involves a loss of oxygen, leading to what is known as an anaerobic environment. The bacteria must survive in both the presence and the lack of oxygen to keep the momentum going.

Even if the bacteria can survive all these conditions, they still have to prove they can stick around, if only for a few days. To do this, they must be able to adhere to the various human cell types along the gastrointestinal tract. Depending on the location, the cells have a different appearance, and some are also covered in a protective mucus barrier. Getting a hold on these surfaces can be difficult for a bacterium unless it is equipped with proper microscopic tools, such as mucus-degrading enzymes and extensions to latch on to the surface. Unless a bacterium can show it can attach to most, if not all, of the cellular surfaces, the journey is over.

For human strains, these challenges can be met fairly easily as they have already gone through them inside the child. But the next stage of testing is completely different—it involves introducing potential pathogens into the mix. Because the human body may accumulate a variety of foes, the candidates have to be able to withstand an attack and quite possibly show how they would fight back. Many strains end up giving up at this point, since they are incapable of producing the antimicrobials necessary to actually fend off the attackers.

Any species making it through this stage is most likely going to be an acceptable candidate for a probiotic. Yet there is one more examination needed before they can be taken to the development stage—they have to show that they are not themselves harmful to humans. Even though these species are known to be beneficial, some can form toxic by-products.

This is highly unlikely for bacteria that came from a human, but to ensure safety, this test is a must.

By the time all the tests have been conducted, the list of candidates may have shrunk from dozens to maybe one or two. These remaining species are the most viable option for probiotic development. They are also far removed from the feces from which they came. They are no longer human strains but laboratory ones. So while seeing "human origin" on a box or bottle may give pause, there's little to worry about. There is more than enough separation to set the mind at ease.

HERE FOR A GOOD TIME, NOT A LONG TIME Even the best parties have to end sometime. The food and drinks run out, moods change, and sometimes even fights break out. It's time to leave.

It's like that for probiotics too. When they are ingested (particularly with food) and reach the intestines, it's microbial festival time, with billions of bacteria feasting on nutrients, mingling with other species, and keeping the mood in the gut light and bright. But as the nourishment disappears, the banquet turns into a battlefield between the native species and the hard-partying visitors. Though the probiotics have contributed to making the gut a most happy place, they are no longer welcome. They have only two choices: they can hitch a ride out with the feces, or they can die on the spot.

For the most part, probiotic bacteria are short-term residents of the gut. They like environments where food is readily available and they can thrive without any competition. In an environment with up to one hundred trillion other members, this can be difficult, especially when everyone has the same goal. Although we tend to eat on a regular basis, the nutrients we provide are not enough to convince most species to stick around.

This trait of probiotics isn't much discussed. After all, their relatively short-term effects would make them seem far less enticing. But we need to understand how these species work inside us, to appreciate their true benefit and not tout myths.

The most touted of the unrealized benefits is their ability to change the microbial population in the gut. The diversity of the gastrointestinal tract takes a long time to develop and goes through significant changes over the course of a child's infancy. But it usually stabilizes at around one year of age. As a person grows and is exposed to various microbes through diet and lifestyle; diversity increases. Scientific studies have shown that we lose microbial stability around the age of seventy, although no one really knows why.

The only way to disrupt this balance is to change diet and/or geography, to use antibiotics, or to reintroduce a mass population of bacteria (also known as fecal transplantation). In other words, it requires either a long-term change in habit or a massive shift in the microbial population, with trillions either introduced or killed off.

None of these options are possible with probiotics, as they are not concentrated enough to invoke any significant change. Even at a level of one hundred billion, that still represents less than 1 percent of the bacteria in the gut. There are simply not enough of them to make any difference.

But this doesn't take away from their ability to improve health. It just makes the timeline shorter (at most three days).

To continue the benefit, all you need to do is eat probiotic bacteria every day or at least every few days.

At one time, that's how probiotics were delivered—not from bottles on shelves but in naturally fermented foods. Some diets that still call for fermented products suggest they should be taken every few days—which is how long most probiotic species stay in the gut. But very few companies have gone back to the old ways and developed freshly fermented products using probiotic species. Most probiotic producers have taken an easier route and simply placed the bacteria in a pill.

Finding a freshly fermented probiotic is not easy, and most of us will have to resort to using a supplement. Yet in this case, popping a pill shouldn't be regarded as therapy or treatment. Instead, the probiotics should be viewed as part of daily life, like vitamins and mineral supplements. If the natural option isn't available, then just pop a pill a day with a meal, preferably one with some fibre; probiotics just love fibre.

AND NOW—PREBIOTICS *Prebiotic* is a relatively new term in the nutritional field denoting a certain type of food product that we cannot digest but friendly microbes love. If that makes it sound an awful lot like fibre, that's because for the most part, that's all a prebiotic is.

Among the various molecules that make up fibre, some have been shown to provide good nutritional support to the beneficial bugs in our gut. These specific molecules, with names like inulin and, more of a mouthful, fructooligosaccharide, are found in many different foods but can also be provided as supplements. We can take these and know the good bacteria will benefit while those unfriendly species will go without as they wait for a more sugary treat.

Taking prebiotics is an excellent way to increase health, but the best place to find them is not a drugstore but a grocery. The best way to get prebiotics is to eat more fibre-rich foods such as vegetables, fruit, grains, and legumes. They may not have the same concentration of prebiotic molecules as some supplements but they can easily be incorporated into the diet without thinning out the wallet.

NOT FOR EVERYONE You'll never hear me say probiotics are for everybody. Although there are many good reasons to take high doses of these friendly bugs, some people should nonetheless stay away from them.

This is because probiotics as supplements are essentially medication. Yes, the bacteria are natural friends, and yes, they can be found in many fermented products not considered to be medicinal. But the concentration of the bugs in food is not as high as it is in probiotic supplements. Companies have made significant efforts to attain concentrations you would never see naturally. So the products are for the most part pharmacological and should be treated as such, side effects and all.

Those with suppressed immune systems due to pharmaceutical use, age, pregnancy, or chronic diseases such as diabetes and cancer, are the most at risk from possible side effects. Friendly bacteria work constantly with the immune system to improve health. But if immunity is compromised, the bacteria could cause trouble as they too can get into the bloodstream. Having large amounts of bacteria in the blood—no matter if they are friends or foes—is bad. It is called sepsis and it can kill.

Sepsis can also occur when the integrity of the intestines is compromised. When an individual undergoes a resection of the colon or other part of the intestines, probiotics could potentially become a problem. That's because these bacteria are beneficial only when they are not in the bloodstream. The intake of bacteria should be avoided unless specified by a medical professional.

The final reason not to take a probiotic has to do with the method by which it was grown. Many probiotic species are grown not in a lab but in their natural habitat, which is usually milk. Anyone with intolerance or allergy to milk would be best to avoid products with his kind of bacterium. If milk was used to make the product, it should be listed in the ingredients. Some companies will also explicitly state that they've used milk in the manufacturing process.

THE NATURAL PHARMACEUTICAL No matter what natural product we consume, the priority is safety. For natural foods and beverages, many countries have regulations to ensure that anyone who eats or drinks these products will be safe from illness or other toxic exposure. But when it comes to probiotics, safety is a major challenge as these health benefactors can themselves be a threat.

The majority of probiotic species are generally believed to cause no harm as they occur naturally in both fermented foods and the gastrointestinal tract. But there are rare instances in which bacteria thought to be safe have caused illness. The most efficient way to achieve a proper level of confidence is to treat probiotic supplements not as the natural source of health they are touted to be. Instead, they need to be regarded in the same light as pharmaceutical products such as drugs and vaccines.

Before any probiotic can be produced commercially, it must go through a variety of clinical trials to prove it is safe to consume and does indeed provide benefits. Every trial is conducted in essentially the same way. Several random people are gathered together and separated into two groups. One group is given the microbes, the other a placebo. This goes on for a period of weeks to months, and then the two populations are compared to determine if there were any changes in their health.

The first phase is to determine whether the bacteria can be tolerated by the body, and if any side effects occur. It's the simplest of the trials, as the goal is to ensure that the bacterial population at the specific concentration won't do anything bad to a person. Usually, this stage passes without incident, although there have been instances in which serious problems occurred. One of the worst is sepsis, which I've already mentioned is a potentially fatal condition. Without immediate antibiotic treatment, the chances for recovery are slim.

These safety trials help to identify who can take probiotics. Thanks to these studies, we already know that people in intensive care units, the very young, those with compromised immune systems, and those who have recently undergone surgery should not ingest these bacteria. The trials also reveal the one aspect of probiotics never mentioned outside of the clinical environment: bacterial behaviour may be completely known in the lab and even in animals, but when those same bacteria are placed in humans, they might simply run amok.

Once safety has been established, the next stage of testing involves determining whether the probiotic actually does any good. The same protocol is applied, but the testing becomes more rigorous. Depending on the parameters, various samples, ranging from fecal matter to blood, will be taken from the participants. Back at the lab, testing is performed to determine whether any benefits can be seen. This could mean looking for bacterial by-products in the fecal matter or searching for signs of immunological calm in the blood.

The threshold for benefit is quite high. The bacteria need to perform effectively for those taking it and also outperform those taking the placebo. Many probiotics end up losing this challenge, thereby halting the process.

If a probiotic makes it past this hurdle, then an application for government approval can be made and eventually the product can be sold on the market. Some companies, however, choose to take a further step towards proving the benefits of their product: they give it to a large audience and wait to see whether a drop in the rate of illness occurs. This is known as an open label trial, and it can help to increase the evidence supporting a particular claim, particularly when it comes to the prevention of illness in a human population.

By the time a probiotic makes it to market, it has gone through a multi-year journey to demonstrate it is safe, has proven benefits, and is a suitable option for improving various aspects of health. The process has also probably cost millions of dollars. From a medical and pharmaceutical perspective, this effort is worthwhile. From a natural perspective, however, there is a sense of irony. These highly regulated products come from rather wild origins and haven't changed all that much in the process. Put another way, the probiotics on the shelf may be no different from those found in raw, unregulated fermented foods; they probably give the same benefit too.

FOR ORAL USE ONLY The growing popularity of probiotics has led to a vast array of products, some of which are intended for areas other than the mouth or the gut. Probiotic skin creams have become common, as have probiotic soaps and shampoos. Even probiotic air cleaners are making their way into stores. While these products may be of benefit, they don't fit the current definition of a probiotic. A probiotic is solely intended to be taken into the body orally. It stays in the body only a short while; during which the bacteria perform one or several tasks known to help make our lives better. That's it.

If you happen to come across one of these non-ingestible products claiming to contain probiotics, you should take a pass.

SOMETHING TO CHEW ON We've seen many attempts to cash in on the popularity of probiotics. Pretentious products have included probiotic cookies, probiotic candies, and even probiotic drinking straws. But while those products offer no real benefits, one odd invention may have staying power: probiotic chewing gum. The concept may seem strange, as gum is usually useful only in freshening breath and in some cases whitening teeth. But some companies have placed probiotic species inside these sticks and pieces and put them to the test. Not only have they proven helpful in promoting good bacteria in the mouth, but they can even help prevent the onset of cavities and other oral diseases.

The trick comes in the way we chew gum. As soon as we take a bite, our mouths are filled with saliva in response to the introduction of food. At the same time, the bacteria in the gum are quickly released from the matrix. They are too small for the teeth to harm them. The rush of fluid picks up the microbes and disperses them throughout the mouth, bringing them into contact with the teeth, the gums, and the cheeks. At this point, depending on the force of the chewing and the sloshing around, there may be an opportunity for the bugs to grab hold of a surface and come to rest.

Once here, the real magic happens. The bacteria first look for any foes in the mouth and engage them for the territory. The foes are simply not ready for the attack and end up dying off. As the process continues, the balance of the microbial population of the entire cavity changes so more good species are present. Once the area is settled, some members of the group will head out in search of other places to colonize. With the help of saliva, they can travel to a wide variety of areas, including the sinuses, the tonsils, and even the lungs.

Probiotic chewing gum can also help to freshen your breath. Halitosis—bad breath—is usually caused by unhealthy microbial species producing stinky chemicals as they eat away at the teeth and gums. But when the good bugs are in place, they have mainly odourless by-products. As their population grows, a more natural scent will end up emanating from the mouth.

There are some limits to all this goodness: the bugs cannot reduce the overall microbial concentration of the mouth, nor can they remove plaque. That's where brushing and flossing are needed. So while probiotic chewing gum may help improve oral health, it is by no means an excuse to forgo your usual dental hygiene procedures.

DISEASE DETECTIVES Doctors have a new diagnostic tool in the form of bacteria. So-called diagnosing probiotics are species that have been modified at the genetic level so they produce extra proteins. When these molecules come into contact with a known disease marker like a cancer cell, they change colour. When excreted in the urine, the colourful markers clearly signal the presence of troubles.

The process is relatively simple and non-invasive. All that's needed is to ingest the bacteria in high enough amounts—in the billions. Once in the gut, they can produce and send these specialized molecules into bloodstream to hunt down the disease of interest. When they find their target, they change in a way that can be detected later on in the urine or the fecal matter.

Apart from offering a rapid and accurate diagnosis, a probiotic disease detective may also be used to monitor the progress of chronic diseases and the success (or lack thereof) of treatment. Instead of having to visit the doctor's office or the laboratory for regular poking and probing, you would simply have to take a probiotic pill and look for changes in the urine.

PROBIOTICS 2.0: PHAGEBIOTICS The presence of pathogens can make human life can feel risky. But our chances of catching a deadly infection are actually pretty small, especially when compared to what bacteria face from the viruses that prey on *them*. Bacteria, in fact, live with the constant threat of not only getting sick but being killed.

These highly hostile viruses are called bacteriophages—or phages, for short. Once one of them encounters an unlucky bacteria, things get ugly fast and death is almost certain. The process is vicious. A phage comes into contact with a bacterium and then sends a piece of genetic material into the cell. This small invader is a code for the production of proteins, and it takes over the cell's manufacturing process. Soon, the entire cell is committed to making more viruses.

When the phage population gets too large, the bacterial cell bursts and releases those viruses into the body to find new cells to infect. The process is quick and can destroy a bacterial population in a matter of days. But the viruses can survive only as long as there are bacteria through which to multiply; they are limited to one or a few species and cannot infect human cells.

There are thousands of different types of phages, and each one has a unique ability to focus on one or more bacterial species. Many of these targets are bacteria we consider to be foes. This makes these viruses a great potential treatment for infections caused by bad bacteria, as well as a means to control their overgrowth. We can culture specific viruses and then give them to a person, knowing that they will quickly destroy the harmful bacteria.

Using a virus to combat troublesome bacteria is a developing idea, but it seems to have the potential to usher in the next generation of probiotics, called phagebiotics. Although we are just beginning to harness the power of these viruses, we can see a number of useful purposes. We can add them to products such as foods and cosmetics to keep these items safe for human consumption. We can also use them in medicine to target bacteria, especially those already resistant to antibiotics. But the real value comes in using the viruses as part of the diet to limit the number of bacterial foes and maintain, if not expand, the amount of good bacteria residing in our guts.

Much like probiotics, phagebiotics could be added to various foods, particularly fermented products with an already high concentration of good bacteria. They could be used as additives to food products such as fibre. Or they could easily be mass-produced and put into a pill for routine consumption. In all cases, our bodies would welcome these small friends and allow them to feast on our foes.

But there is a significant hurdle in getting phagebiotics onto store shelves. Because bacteria are constantly under threat, they have developed a mechanism to fight back. It's essentially a bacterial immune system. Once a virus's genetic material finds its way into a cell, the bacterium may actually intercept it before it has a chance to start making proteins. At this point, the genetic material could be used to develop a defence strategy. If another virus happens to come into the cell, the new mechanism is used and the genetic material is not only neutralized but destroyed. It's an effective strategy that seems to be able to outlast the virus's ability to infect. When phages and bacteria are allowed to commingle over a period of months, the bacteria inevitably win out, forcing the phages to die off.

The key to phagebiotics is to ensure that bacteria don't gain resistance to them. We therefore have to be prudent about their use, much as we are with antibiotics. This means they are unlikely to come into widespread use any time soon. But once we've worked out how to protect phages from the bacterial immune response, this valuable and abundant resource can be turned into a product that will help to keep our bodies and our friendly bacterial populations happy and healthy.

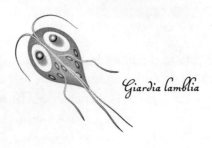

Giardia lamblia

6. HEALTH

FLU BUGS Every fall and winter, we do our best to avoid the collection of viruses known to cause colds and flu. Hygiene plays an important role in keeping us healthy, but so do our friendly bacteria. Their contribution is less about strengthening our immunity and more about maintaining balance and focusing those defence forces on the right target.

Colds and flu put our bodies through the wringer. These viruses have a keen ability not only to multiply and initiate a strong invasion but to frustrate the immunological troops. When this happens, the immune system goes on a rampage. Targeted attacks are abandoned as they prove ineffective. Instead, the immune system creates a killing field.

The result may be the destruction of the virus, but human cells can become collateral damage. This may end up making things worse, as symptoms tend to last far longer and a two- or three-day cold can turn into a three-week nightmare, complete with conditions ranging from pneumonia to asthma.

Controlling the immune response is not easy. When the system has decided to attack, there are only a few ways to halt its action. Aspirin and other non-steroidal anti-inflammatory drugs are excellent at keeping the immune system from over-reacting. They block a certain immunological function called inflammation from taking over the entire operation. This is a very good thing, as most damage comes from the effects of inflammatory attacks. Yet these drugs can help only once the inflammation is detected (usually that means a fever or a nasty cough). Damage can occur even before the inflammation is noticed.

Bacteria know about inflammation all too well. Some worsen the situation by fighting back and attempting to force the body to self-destruct. Others calm the torment through an indirect interaction with the immune response. The bacteria send signals to human cells in the gastrointestinal tract, reassuring them that all is under control and there is no need to overreact. This in turns leads the immune to calmly control inflammation throughout the body. Many friendly microbes carry out this beneficial form of control, but when it comes to reducing the damage from flus and colds, the best species seem to be the probiotics.

Probiotics end inflammation by sending the immune system a "stand at ease" signal rather raising an alarm. If there's enough bacteria in the gut, the signal spreads to other areas, including the respiratory tract.

And there's another added benefit to probiotics: they are able to focus the immune response to go after the viruses. This particular command is not really designed to keep us healthy;

the probiotics simply want to keep the immune system preoccupied while they grow. But the end result is more protection for us because visiting viruses are given less chance to settle in and begin an infection. This also prevents the immune system from overreacting and accidentally taking out the probiotic bacteria themselves while the body tries to fight.

But probiotics are not guaranteed to stave off colds and flu. At best, they reduce your chances of contracting these viruses by about half. This may be helpful to those most at risk, such as children and the elderly. They already have weaker immune systems and so they tend to experience more dramatic effects from infection. To further reduce your odds of infection, combine probiotics with good hygiene: handwashing, hand sanitizers, and those stylish masks known as scarves. But even if you get ill, these friendly bacteria can still help you to recover faster and prevent complications meanwhile.

ALLEVIATING ALLERGIES Researchers have found that a lack of exposure to a variety of microbial species early in life tends to increase the chances of an allergic response. Why? No one is sure, but it seems to have something to do with the way allergies come about.

An allergic response is akin to having an infection. When one or more small molecules, called allergens, get into the body, they are recognized as foes. This triggers an immune response to get rid of them. But this is where the circumstances change. Instead of the usual process of seeking out and destroying the foe, the body guards against the threat in one of two different ways. The first is through dilution, which is why allergy sufferers experience that sudden rush of fluid in the eyes and sinuses. The other is through constriction, in which the body closes up shop to keep the offending object away. It's why asthma sufferers cannot breathe.

The source of the problem is the defence forces' inability to recognize the difference between a harmless foreign object and a hostile one. This has less to do with the location of the symptoms than with the diversity of the bacteria in the gut. When there is a large variety, the body's reactions seem to be less intense. But if the balance is off—if the diversity is lower or skewed towards foes over friends—the body will react more strongly, worsening the symptoms. The connection between allergies and the balance of bacteria in the gut is still not entirely sorted, but this hasn't stopped researchers from using friendly bacteria to alleviate the condition.

Take a peanut allergy, for example. It's a vicious attack in which the entire body undergoes constriction, with some-times fatal results. The procedure for curing peanut allergies is called immunotherapy and should only be done by a profes-sional such as an allergist and never be tried at home.

This highly controlled process begins with the patient being exposed to a very small amount so as not to initiate a reaction. The dose is increased over six months. The goal is to train the immune system to tolerate the proteins contained in the pea-nut. As the body becomes used to exposure, immune defences will turn off their attacks. If treatment is fully successful, the person can be free of an allergy within two years.

The success rate for immunotherapy varies but is usu-ally better than 50 percent. Seven out of every ten people undergoing treatment are given the all-clear. This is a pretty good outcome but there is a way to make it even better: by adding probiotics to the mix.

The protocol is similar to immunotherapy with one exception: with each dose, twenty billion probiotic bacteria are also provided. Within a year, up to 85 percent of sufferers find their allergy problems disappear.

The key to success lies in the probiotic's potential to change the overall nature of the immune response. Including the friendly bacteria with the peanut can essentially reprogram the immune system so it responds less severely or not at all. This treatment is both more effective and faster-acting than traditional therapy.

The practice of using probiotics to alleviate allergies has yet to gain wide acceptance, but I'm sure it will. Taking regular doses of probiotics during allergy season can help to reduce the levels of attack and might even obviate the need for medications like antihistamines. It won't cure you of the problem, but it may help to stave off those frustrating symptoms.

FIX THE FLUSH Our intestines are sensitive to the presence of sugar and salt, and notice when their concentrations rise. If these nutrients cannot be broken down, there is nothing to be done but flush out the offending molecules as fast as possible. The intestines flood the gut with water and push the entire mix towards the exit. We call this diarrhea.

But the runs can be slowed down. Probiotic species such as the lactic-acid bacteria love to gobble up these offending molecules, effectively preventing the gut cells from reacting to them. This is particularly helpful in cases of intolerance, where even the slightest dose of a molecule such as lactose can lead to gas, bloating, and diarrhea. When probiotics are consumed regularly, particularly in the form of yogurt, the lactose-intolerant are sometimes able to accept milk and other dairy products.

Diarrhea is also the body's answer to toxins. (Some laxatives take advantage of that fact by using toxic ingredients to trigger this response.) Microbial toxins vary in shape, size, and function, but all have the ability to alter water retention. Cholera forms a particularly vicious toxin that not only destroys the inner lining of the intestines but also alters the way the cells use water. Our bodies are forced to let go of their supply, causing a massive flood of water into the intestines. Without treatment, sufferers can die from dehydration.

Gut bacteria offer some protection against diarrhea from these chemicals and toxins. They trap the molecules so they are unable to come in contact with the intestinal wall. The addition of probiotics strengthens the microbial population's ability to ward off the effects of toxins and slows the delivery of certain pharmaceutical agents. These good bacteria also obviate the need for laxatives and other stool-loosening drugs. As natural regulators of digestion, they can maintain a better flow and prevent constipation.

The final cause of diarrhea is a microbial imbalance in the intestines favouring potential pathogens. Sometimes, as in the case of food or water poisoning, this is caused by infection. It can also occur after an antibiotic treatment, which wipes out certain parts of the microbial population and leads to imbalance.

These are short-term issues, but some people suffer longer-term problems caused by a lack of gut diversity, including irritable bowel syndrome, inflammatory bowel disease, ulcerative colitis, Crohn's disease, and *C. difficile* infection. In all these cases, the body's internal biological environment becomes less harmonious and forces the intestinal cells to add water to keep things cool, which results in the watery waste.

Probiotics can combat the bad bugs using an assortment of antimicrobial chemicals, killing them and lowering their numbers. But their true value comes from raising the number of healthy bacteria in the gut. This improves diversity and also sends chemical signals to other species to work as a team rather than as individuals. The gastrointestinal cells are then better protected.

But prevention is not cure, and probiotics are not going to stop an already occurring problem. They can supplement other treatments, but can only do so much on their own. That being said, probiotics can limit the damage that's sometimes caused by taking antibiotics.

GOOD FOR THE GUT Hidden deep within the colon is a bacterium called *Faecalibacterium prausnitzii*. It has always been in human feces, but it wasn't discovered until 1922 by a researcher named Prausnitz (Carl Prausnitz, to be exact). This species is one of the most important for our gut health and deserves a catchier name: the probiotic fixer-upper. The bacterium is notable less for its presence in and more for its absence from the gut in those who suffer from various diseases.

Millions of people around the world have one of three debilitating gastrointestinal conditions: inflammatory bowel disease, colitis, or Crohn's disease. For these people, the gut is a constant menace, bringing pain, diarrhea, and blood in the feces. It can be a lifelong affliction, as for many, there is simply no cure.

Examination of the feces of sufferers reveals that *Faecalibacterium prausnitzii* is either missing entirely or exists in numbers far below those in healthy people. This doesn't necessarily mean that a lack of this microbe is the sole reason for people's troubles. But if you add it to the gut, something amazing occurs: the symptoms of the disease are reduced (at least in mice—this has yet to be tested in humans). The condition is not reversed, but there is a dramatic improvement.

Why? It's all to do with the cause of these ailments: the immune system. The sufferer's defences mistake what should be friends for foes, and the process spirals into constant battles without hope of a cure. This all changes when Prausnitz's bug is in the colon. The bacterium produces long sugar molecules, which coat the gut lining. The intestinal cells recognize the protective layer and signal the rest of the area to reduce any attacks. With enough of these bacteria in place, the good times can become systemic.

But that's only the beginning. The bacteria also make a number of molecules known for their ability to take on a raging immune system and calm it down. Some are direct anti-inflammatory chemicals akin to Aspirin. Others are designed to signal the cells responsible for keeping the troops in line. They regulate and suppress the activity so the gut can relax. Apart from these overt actions, the bacteria also produce several subtle chemical signals to help soothe the cells in the gut. The effect is so great in the mice that the autoimmune conditions can be successfully treated.

Of course, controlling the levels of bacteria in the gut is not always easy, and so researchers are working hard to develop a *prausnitzii* probiotic supplement. Unfortunately, there's a slight problem with this species that makes that path almost impassable: the bacterium hates oxygen. Unless it's covered in antioxidant molecules, it cannot survive in normal air. It would take an incredible effort to keep oxygen from entering a manufacturing environment, and it would make the supplement so expensive that no one would be able to afford it.

All is not lost. The bacterium is, after all, already inside us. All we need do to make it thrive is to feed it properly, by eating more fibre ourselves. Fibre molecules are the perfect food for this bacterium, and it will thank you immensely for passing them on. And to make it really happy, choose products that contain inulin such as bananas, asparagus, garlic, and chicory root.

GUT REACTION There is a condition called small intestinal bacterial overgrowth, or SIBO. It causes several problems, including intestinal blockage, weight loss, malnutrition, cirrhosis, constant pain, and the inability to fight off infections.

The source of the problem is an overabundance of bacteria—usually pathobionts—in the small intestine. Normally, there are about one to ten thousand cells in every millilitre of intestinal fluid. In people with SIBO, there are up to a hundred times that number—well over one hundred thousand for every millilitre. The bacteria also tend to clump up in colonies, attaching themselves to the intestinal wall and blocking the body's ability to absorb nutrients. This triggers the immune defences into trying to reduce their number to normal levels. But being outnumbered, they can do little to help.

SIBO is a complex condition, but it is directly related to diet and excessive use of alcohol and drugs, including those used to control heartburn. Although we know how to prevent the condition, following through can be difficult and many of us simply cannot make all the required lifestyle adjustments.

There is a medicinal treatment in the form of antibiotics taken over a short period, but it is not always that effective, and relapses can occur without those needed changes in lifestyle—changes that don't usually happen during the relatively short prescription time. But the situation can be helped with the introduction of probiotic species. They have the ability to work within the small intestine and help knock out the unfriendly bacteria. During their short stay, these beneficial bugs not only fight the invaders but also bolster the defences to ensure a more successful attack. Add in some prebiotics like fibre and the therapy is even more effective. Once the prescription is done, sufferers can continue using probiotics and prebiotics to keep the pathogens from re-emerging. The supplements are not a cure, but they can maintain microbial balance while other lifestyle changes are made. Given the consequences of having SIBO, it's worth trying anything that keeps away the pain.

GERMS THAT HATE THE GYM I love watching television and loathe going to the gym. I'm sure I'm not alone. I try to justify this preference by saying that I'm trying to help my microbes stay calm and avoid the perils that come with human physical exertion. But no one ever seems to buy it.

It's true, though, that some microbes don't much like it when we exercise. And if we exercise regularly, those species will jump ship and go in search of a lazier environment. This shift in microbial population occurs as a result of physiological stress. During exercise, we are wilfully depleting the body of oxygen and energy. The response to this is inflammation in the physical context—think "feel the burn"—and also the immunological one. When this happens, any unfriendly bacterium may end up having to deal with an unexpected attack by a group of immune cells called neutrophils.

Neutrophils eat and destroy uninvited visitors. When they come across a bacterium or virus that is unknown, or at least appears to be unhelpful, they ingest it and then break it down once inside the cell. Then they release toxic chemicals into the area to weaken any threatening entities that remain. These changes make the gut lining more permeable, allowing water to escape into the intestinal cavity. Our bowels become loose, and we may experience diarrhea, cramping, and nausea.

Our gut microbes are wary of neutrophils and take their own defensive actions when they encounter them. For some foes, the evasive manoeuvre is simple—catch a ride with the diarrhea and get out of town. For others, it means sending toxins into the body, which worsens the overall effect of the inflammation. The entire gastrointestinal tract can become a war zone in which the mightiest survive and the weakest fail.

Within a few hours after exercising, everything returns to normal. For our bodies, this type of short-term cycle is actually good, as it keeps our immune systems active and ready for any real battles to come. For the bacteria, continued exercise can lead to a more harmonious balance of friends, bystanders, and of course, foes. This in turn leads to a change in our dietary cravings, such that we lose that addiction to sugar and fats and instead seek out more nutritious sources of carbohydrates, as well as proteins for the energy they provide.

There is one caveat to the effects of exercise: if you choose not to change your diet and continue to seek out foods rich in simple sugars and high in fat, the shift in your gut population will be minimal at best and the lost benefits will discourage you from attending the next gym session. With each increase in unhealthy foods, the couch looks just that more inviting.

BACTERIA NEED BEAUTY SLEEP TOO Our bodies depend on a regular cycle of movement and rest called a circadian rhythm. It's a biological dependency on spans of light and dark. When the earth is in sunlight, our brains force us to be active. At night, we are compelled to slow down. If we interfere with this biological clock, we may become more prone to illness and even reduce our life expectancy.

This same need for regular movement and rest can be found in our microbial population, particularly species found in the gut. Bacteria do not sleep, but they do rest. Inside us, they tend to follow a schedule similar to ours. They spend the daytime hours eating, metabolizing, and multiplying, and during the night they become less active and focus on maintenance, including repair and detoxification. This change can be seen in the numbers of each type present. During the daytime, some species, like the lactic-acid bacteria, can increase as much as 15 percent over amounts seen at night. They multiply quickly when being fed during daylight hours, and die nearly as quickly in the absence of nutrients during sleep.

Bacteria tell the time by the presence of nutrients, such as those provided by our own meals. When we eat, they know it's time for action. Between our meals, they can relax. The alignment of rhythms not only keeps our bacteria happy but also improves our overall metabolism, helping us to stay in shape.

If we maintain our three square meals and get seven to eight hours of sleep at night, the bacteria work alongside us to keep everything balanced and calm. But if a person messes with the rhythm, it means trouble for the microbes. The species needing that regular nutrient/starvation rhythm begin to suffer. A blip in the rhythm is akin to a massive shift in their environment, and confusion sets in. They seem unable to synchronize. A once-friendly home is seemingly no longer hospitable. This is when that 15 percent change really takes effect. The bacteria that are switched on will overgrow and become too populous, while those that are switched off will decrease in number significantly and may disappear altogether. The result is a loss of diversity, known as dysbiosis.

In the sleepless, dysbiosis can bring about a change in metabolism. The body can use only so many nutrients in food; the rest have to be broken down by the bacteria in the gut.

Without them, we lose out on the potential for balanced sugar levels and efficient fat burning. We feel lethargic and may put on weight. Over time, we may develop mood disorders, lowered brain function, diabetes, or cardiovascular disease.

The way to reverse this trend is simply to get more sleep. If for whatever reason that's not possible, you can still trick your gut bacteria into thinking everything is going well. You do this by dividing your usual food intake over the course of a day and spread it out into more meals during both day and night hours. This cycle of feeding and starvation will balance the numbers of bacteria. This allows you to travel and enjoy the sights while keeping your gut believing everything is copacetic.

DEALING WITH DIABETES Diabetes is a growing problem. People around the world are suffering from an inability to metabolize glucose, the most important nutrient for life. It's become a crisis in many countries, and medical professionals are doing their best to stop it. But controlling diabetes seems to be a greater challenge than we ever expected.

Normally, the insulin hormone is needed to get glucose into the cells. If the cells are unable to get that nutrition (a condition known as insulin resistance), they can starve and end up dying. The consequences of diabetes can include the loss of proper blood flow to areas of the body, reduced kidney function, and even the need to amputate a limb. There are actually three very different faces of this disease, each with a different cause and a different treatment.

Type 1 diabetes is a lifelong condition in which our body no longer produces insulin because the cells manufacturing the hormone are self-destructing. There is no cure for this particular form and the only treatment is regular insulin injections. Type 2 diabetes is entirely different, in that the body still produces insulin but the cells of the body have somehow become resistant to it. This type of diabetes is not only preventable but may be reversible, thanks in part to microbes. Type 3 diabetes affects only pregnant women and is called gestational diabetes. Around the beginning of the third trimester, the hormones produced during pregnancy begin to rise, and one of them, cortisol, causes the cells to resist insulin in the same way they do with Type 2 diabetes. But this is a short-term problem and many mothers return to normal after birth (though their risk for diabetes later in life is higher).

Microbes have an important role to play in Type 2 and gestational diabetes, in both prevention and, in some cases, control. The link between the bacteria in the gut and the onset of insulin resistance is quite complex and involves the metabolic, immune, and hormonal systems. Yet there appears to be one molecule engaged in disrupting all these systems. It's called the tumour necrosis factor alpha (TNF-α), and it's important in the prevention of cancers. It also plays a significant role in making your body aware of a potential threat and ensuring it is prepared for battle. When any part of the body is in distress, TNF-α signals the rest of the surrounding cells to focus their attention on the injured or stressed area. When the situation resolves, TNF-α levels drop and everything returns to normal.

When TNF-α signals the metabolic system, it begins to slow down its use of sugar. (There is more energy in fat, which has nine calories per gram, than sugar, which has only four.) As a result, insulin's role is reduced primarily to conserving energy for later. After all, who knows how much energy will be used up taking care of that remote problem?

Microbial imbalance in your gut isn't always a short-term issue and may go on for weeks, months, and even years. Add a poor diet high in sugar and fat and the situation worsens. Only certain microbial species will thrive, while others will lose out because of a lack of nutrients or competition for the incomplete diet. As these microbes battle it out, the intestinal cells continue to be stressed and on their guard. This means they're continually producing TNF-α and slowing down metabolism. The effect is so powerful that insulin resistance eventually sets in, as does Type 2 diabetes.

Probiotics can help to resolve the problem. They do this by sending your intestinal cells helpful signals to ensure that the area is safe from internal competition, resulting in a reduction in TNF-α. That being said, probiotics cannot cure diabetes or lead to long-lasting changes in the gut. But they can help smooth the path to health while changes in lifestyle and diet do their work.

COMBATING CHRONIC DISEASE Solving the riddle of chronic disease is incredibly difficult. Prescription medicines are usually the only way to move forward. But there may be a new way to treat these ailments using microbes.

Like humans, several species of bacteria have an immune system. It's nowhere near as complex as ours, but it can protect against a certain type of virus, called a bacteriophage. This is quite an important job, given that when a phage attacks, death is usually inevitable.

The microbial version of immunity is known as clustered regularly interspaced short palindromic repeats—but we can just call it CRISPR. Within the genetic material of the bug, there are areas capable of mutating in such a way that they produce molecules to block the phage attack. This mutation gives the cell a form of microbial memory, and is permanent. If another virus attempts to invade the bacterium or any of its future generations, it will fail.

Memory after mutation is what makes CRISPRs so valuable to medicine. The bacteria can be genetically engineered so they make not an anti-phage molecule but a medicinal one, such as a drug or an anti-cancer agent. Once the change has been established, every bacterium grown subsequently will be a tiny therapeutic machine, pumping out the drug of choice.

But production is only one part of the solution. Most chronic conditions require several doses of medication a day by way of pills or injections. With a bacterium, that can be reduced to one dose a day or even one every few days. Also, the treatment could be more efficient, since bacteria are already excellent at sending molecules into the gut, the tissues, and the bloodstream.

This kind of microbial medicine is years away. After all, CRISPRs were only found at the beginning of this century. But because this immune system is so simple in design and so easily engineered, it holds promise as an achievable and highly desirable form of treatment.

HEART-HELPER Cardiovascular disease is one of the most common causes of death and is usually associated with a rise in cholesterol. Our bodies need this fatty substance to keep cells intact. But we need only a small amount at any given time. Any extra is stored within the blood vessels, causing a buildup known as plaque. This accumulation can restrict blood flow and lead to a heart attack.

Getting rid of excess cholesterol can be achieved in many ways, including proper diet and exercise. There are pharmacological options as well. Some of the most common are drugs known as statins. These molecules block the formation of cholesterol so we don't have to worry about reaching those high levels. There are several on the market, but they all have a common and unlikely ancestor: fungi.

In the wild, fungi are always in a battle against invading bacteria and yeast. To ensure victory, they produce several antimicrobial chemicals to kill any foe. Each one has a specific purpose. Antibiotics target metabolism. Antimicrobial peptides break down the external wall of the bacterium effectively killing the enemy.

Statins, on the other hand, act as a Trojan Horse. Once they enter a cell, the molecules target a range of microbial activities, forcing the cell to stall growth, and halt the formation of colonies. Eventually, the bacteria die off.

One of the mechanisms behind this statin attack prevents fat formation. Fats are needed to maintain the integrity of the microbial cell; without them, the cell is doomed.

For humans, statins are a medical miracle. Some are still based on the original structures found in nature, while others have been modified into more effective forms. Yet no matter which one is used, the outcome is always the same: cholesterol levels decrease, as do the chances for cardiovascular problems.

As for your gut microbes, they are not greatly affected by the drug. Some will die, of course, but there will not be a significant change in the number or diversity.

DOWN WITH CHOLESTEROL Reducing cholesterol takes work. You have to lower bad fat intake by giving up some of the tastiest foods, including many meats and processed foods. It's a reasonable health request, but adhering to it can be difficult. There may be a way to reduce cholesterol, however, not by removing foods but by including bacteria in the daily mix. Some species, such as *Lactobacillus reuteri*, have the ability to lower cholesterol by short-circuiting the process by which it is made.

The process is complicated—the bacteria don't simply make a cholesterol-lowering chemical. Instead, *L. reuteri* focuses on bile, which is made from cholesterol. In the gut, the bacteria grab hold of the bile and break it down using a collection of enzymes. This in turn signals the liver that more bile is needed. Cholesterol is taken in, converted to bile, and then sent to the intestines. As a result, blood cholesterol goes down.

Most probiotic species have these enzymes and can use them to help maintain cholesterol levels. But certain types of *L. reuteri* are high producers of the bile degraders and over time can actually reduce the levels of cholesterol in the body. These bacteria have been isolated and studied both in the lab and in clinical trials, and they've shown how they can improve heart health. All that's required is to take in ten billion bacteria with every meal. When combined with a healthy diet, these bacteria will ensure that the levels of cholesterol drop in as little as a few weeks. Keep the practice for up to nine weeks and the levels drop by as much as 10 percent. That number is music to a cardiologist's ear.

PATH OF YEAST RESISTANCE In natural health stores, whole sections are devoted to the prevention and treatment of an infection called *Candida*. The name is Latin for "white" and refers to a collection of yeasts known occasionally to torment humans.

These species are found in about three-quarters of us and normally are just part of our overall microbial population. But they are extremely opportunistic and can take advantage of even the slightest opening to overgrow. When this happens, it can mean trouble for the mouth, the skin, the genitals, and the gut. *Candida* can even infect the blood, causing a life-threatening condition known as candidiasis.

The trouble with *Candida* begins when the yeast realizes it can form biofilms. Yeast is normally a small, round cell, but when those biofilms form, the cells group together into long chains called hyphae. These chains are extremely hard to kill and can overwhelm our natural defences. There is little that can be done without some form of medicinal treatment. Thankfully, these treatments do exist, and many can be found in the products on those health store shelves.

But treatment is never as good as prevention. The key to keeping *Candida* in line lies in our mucus, the natural substance we all make to keep our insides smooth and lubricated. This goopy mix of molecules contains proteins, sugars, and fats— all designed to ensure a smooth passage for the cells. Within the mucus matrix are over a dozen proteins known as mucins. Several of these can be used by bacteria and yeasts as food, but some are simply too difficult to break down. One of these mucins even has the ability to prevent *Candida* from attaching, meaning the first stage of biofilm formation is thwarted.

Unfortunately, this form of mucus is normally found only in the lungs and the stomach, leaving the most common areas unprotected. The only way to keep *Candida* from these areas is to find a way to protect the mucus or kill the yeast before it can cause troubles. This is where bacteria can come into play. Many probiotic species dislike *Candida* and do their best to prevent the yeast from gaining any hold in their territory.

Probiotics work most effectively by hogging all the room in the mucus so *Candida* cannot find a place to settle. The good bugs can quickly cover the entire surface area, forcing *Candida* to either move on or attempt to fight for space. If the yeast attempts to invade, it soon realizes it has made a mistake. Several probiotics have an arsenal of antimicrobials at their disposal. Any sign of an attack will initiate the release of a number of toxic chemicals to kill the yeast in its tracks.

Probiotics also have the ability to signal the body of an impending attack. When they do this, the human cells in the area start to produce their own antimicrobials. These molecules help to defend the area in the short term, buying the defensive troops time to make their way to the area to take on the invader and put an end to its campaign.

The best way to avoid *Candida* is to ensure that there are no places for the yeast to reside. This means keeping the levels of good bacteria high. Probiotic pills will be effective in the gut, but if you want to include your mouth in the coverage, your best options are freshly fermented drinks containing high concentrations of these good bacteria. These won't be found in the anti-*Candida* section of the grocery store, however. You'll have to look for them in a fridge. Probiotics will also indirectly help to keep the human defences strong in the skin and genital regions. But to make sure the yeast won't be able to gain a hold, nothing beats personal hygiene. *Candida* is effectively removed by soap and water, and it can be killed by alcohol sanitizers. Though this latter option may not be the best for the genitals, other regions of the skin will benefit from a good twenty-second rubdown.

WIPE OUT It was long believed that urine is intrinsically sterile. But it seems this is not the case. Both men and women have microbial populations in the bladder made up mostly of friendly bacteria from the skin's genital regions. These species have little to no health impact, because their concentration is quite low (around a few hundred for every millilitre of urine). Compare that to the gut, where the concentration of bacteria may be in the billions per gram of feces.

However, a few bacterial species are known to cause a chronic condition called urge incontinence. Sufferers feel the need to urinate very frequently. These bacteria are normally found in the gut, but they can make their way to the urinary tract and settle in the bladder. Here, they interact with the muscles, leading to a low-level immune response. This prevents a full-blown urinary tract infection but leaves a chronic illness of which the main symptom is that incessant push.

The source of these microbes happens to be the gut, so the most likely route to infection is poor hygiene. One cause is incorrect use of toilet paper: we should never wipe towards the genitals.

HEAVY METAL BREAKDOWN Thanks to the Industrial Revolution, we are frequently at risk of poisoning from heavy metals such as lead, arsenic, cadmium, and mercury. We can get mercury and lead from fish, arsenic from the water, and cadmium from grains. Unless these metals are consumed in high concentrations, the risk is relatively low, but regular exposure to low levels can lead to accumulation in the body and the development of a variety of illnesses, including heart disease, diabetes, lung cancer, kidney failure, and brain degeneration.

Once inside our body, most heavy metals are caught and removed in the urine, but small amounts can evade detection and end up in the blood, organs, and other tissues like muscle. Once there, they can remain for years. The cumulative effect of constant exposure eventually causes the cells to die, and that in turn can cause entire organs to fail.

There is no human mechanism to prevent these metals from getting through the intestinal wall. They are so small they can simply make their way into cells and the blood undetected. The only way to prevent this from happening—other than to avoid exposure in the first place—is to trap the metals so they are held tight until removed in the feces. This is where gut bacteria may play a role in keeping us safe. Thankfully, they are not fond of heavy metals either.

When gut bacteria come into contact with heavy metals, they are at immediate risk of harm. Their small structure makes it easy for even the slightest amount of the metals to cause problems. To ensure their survival, they have developed a number of ways to neutralize the toxins so they cannot cause a premature death.

The most effective method is simply to grab the metal and keep it on the bacterium's outer wall. This is relatively easy because there are no specialized molecules needed. The outer shell of a bacterium already has properties designed to attract and trap elements. Once it has a hold of the metal, it adds chemicals to attract even more atoms. This causes something called a nucleation event, which simply means that the bacteria gather the individual metal ions and group them together into larger clumps. Eventually, each clump becomes so large it falls off and exits the body in feces.

Most environmental bacteria are used to being exposed to heavy metals and have no problem dealing with them. The same goes for bacteria living inside our intestines (although they are less efficient). But some probiotic species are able to deal with both the environment and the human body to accomplish heavy metal breakdown, keeping both them and us safe.

The major players are the lactic-acid bacteria. They have the ability to grab on to metals and keep them in place without letting go. Some will create clumps while others simply hold on to the individual atoms, sometimes for days at a time. This is excellent for us, as these bacteria are around for only a short while. On their way out in the feces, they take the metals with them.

THE CANCER CONNECTION When it comes to cancer, microbes have a bad reputation—and for good reason. Viruses and bacteria have both been identified as cancer-causing pathogens, and their presence can increase the risk of troubles in the future.

The most common culprits are a group of viruses that cause hepatitis, warts and the kissing disease, mononucleosis. These species gain access to a human cell and head straight to the genetic material. When they are there, they cause all sorts of havoc, including changes in the DNA. Normally, when something messes with the DNA, the cell is a goner. But in this case, the opposite happens—the cell starts to reproduce at an extraordinary rate. This process continues until finally a tumour forms.

Certain bacteria can bring about the same result, although they operate somewhat differently from the viruses. Instead of entering the cell and causing a ruckus themselves, they send out molecular messengers to do their dirty work. Once these chemicals contact human cells, they send signals forcing each one into a state of shock. The cell tries to respond by making its own proteins, one of which is a life-extending protein that keeps the cell from dying. This is where the real trouble starts. If the shock never ends, the cell will continue to stay alive. As it reproduces, it creates other cells that will also choose immortality over death.

There is another microbial connection to cancer, although very few people know about it. It turns out that certain microbes may be preventative, as their presence is linked to lower cancer rates. These species may soon have a fundamental role to play in preventing cancers from happening.

The best-known case of microbial prevention of chronic disease involves probiotics and colorectal cancer, one of the five most common cancers worldwide. As the location of the condition happens to be the place with the highest population of bacteria, it would only make sense they would play a role.

When the body has a rich population full of diversity, the odds of a cancer developing are quite low. But as the diversity changes and the bacterial population becomes filled with foes, the environment grows toxic. The risk for trouble rises with each toxic substance released. But this can be resolved through the use of probiotic bacteria. The main species interact with the colon cells to keep them more balanced and resistant to shock. They also help the cells maintain a normal lifespan, so that when each individual cell's time is up, it dies peacefully. The probiotic bacteria even help to stimulate tumour recognition in that area so the human body can keep an eye out for any changes in the way the cells act.

The anti-cancer benefits of probiotics may not be limited to the gastrointestinal tract. Fermented foods and other lactic-acid bacteria—the primary species in probiotics—appear to help reduce the likelihood of breast cancer. The link is based on one of the triggers for the initiation of cancer, an estrogen compound known as estradiol. It's the most common hormone found in women, and it's important in the formation of breast tissue during puberty. But after menopause, too much of this chemical can lead to cellular overgrowth.

Preventing higher levels of estrogen is partly dependent on diet. The liver is the primary remover of the hormone, and it does that by changing the structure so the hormone is no longer active. The estrogen is then mixed with bile and sent to the intestines for removal in the feces. With a healthy and diverse microbial population, this happens without issue. But too many foes in the gut leads to a sinister turn of events: the estrogen molecules are returned to their natural state. They are then reabsorbed and allowed to roam the body yet again. The concentration of estrogen can then increase, as can the risk for troubles in the breast.

This is where probiotics may make a difference. They can help maintain the balance of bacteria, keeping the foes at bay, so estrogen can't return to the body. But probiotics are most effective when two more components are added to the mix. The first is fibre. This indigestible form of sugar has a twofold effect: the probiotics love it and are able to keep their numbers high while it is around, and it also helps to keep the bile flowing so more estradiol is removed from the liver. The second important component is a different type of estrogen called phytoestrogens, which is found in some fruit and vegetables, such as apples, oats, soy, lentils, and licorice. These chemicals can dilute human estradiol and keep levels below the troublesome threshold.

There may be one more potential benefit to probiotics: they may help to prevent tumours at the site. The breast has its own microbial population living just under the skin. These microbes mean no harm and cause no troubles, so they haven't attracted much attention. But expect that to change. Certain species break down estradiol and could be a powerful tool in preventing this path to cancer.

One such species is *Sphingomonas yanoikuyae*, which regards estradiol in the breast as food. Not surprisingly, this species seems to be absent in women suffering from breast cancer.

Of course, this might be a matter of correlation and not causation, but it's an interesting observation that opens the door to a new path for prevention. If proven effective in the lab and in clinical trials, this oddly named microbe might become a key ally in helping women fight one of their greatest health threats.

SMOKE ALARM Prolonged exposure to tobacco smoke can wipe out whole populations of the friendly microbes living in the respiratory tract. But others species, including many that cause respiratory and sinus infections, can detoxify the harsh chemicals, survive, and stick around.

As the diversity in the respiratory tract drops, these species take over the area and cause disease. This is double trouble for smokers, as it weakens the immune system in the lungs, the sinuses, the mouth, and even the ears. The overall result is an increase in the likelihood of infections in these areas.

But direct exposure isn't the only problem. Though the immune system is depressed in the areas where smoke is inhaled, the rest of the body experiences an overcompensation of immune function in the form of inflammation. This essentially means our defence forces go into hyperactive mode. Bacteria normally considered to be friendly are left vulnerable to attack and either escape or die. Smoking reduces diversity in the gastrointestinal tract and allows species known to cause infection and inflammation to thrive. In women smokers, a reduction in the number of good bacteria can also lead to gynecological problems, such as a higher risk of yeast infections.

If smoking is a one-off, such as a celebratory cigar, the body will quickly recover from the inflammation. But over time, without reduction of tobacco use, there may come about a vicious cycle. Because the immune system is waging a losing battle against pathogens in the lungs, the mouth, and the gastrointestinal tract, it loses its focus on other bodily systems and cells. The lack of balance and tolerance in those areas can then lead to other problems. Metabolism is affected as the body continues to seek out energy to maintain the battle. Though this may mean a thinner waist, it can also lead to problems with insulin resistance and reduced liver function. Because inflammation is indiscriminate, many human cells will fall victim to the attack, primarily in the respiratory tract. When this occurs, the remaining cells may gird themselves for battle and alter their physiological structure, bringing about thickening of the arteries and heart muscles. For some cells, inflammation sparks mutations at the genetic level, including cancerous ones leading to the development of tumours.

Quitting smoking is the only answer. Within forty-eight hours, the body begins to slow down the inflammation. This in turn can help the microbes return to a more balanced state. But people who quit smoking may find they gain weight. This is in part because of the reappearance of those good bacteria. When they return, the body is overwhelmed and doesn't know what to do with the energy it once burned in battling inflammation. So it stores it as fat.

DRINKING PROBLEMS We are exposed to hundreds of different chemicals in our diet, but one of the most important in terms of human health is ethanol. This molecule is found in all alcoholic beverages, and as such has been a part of human civilization for millennia.

Microbes have a love–hate relationship with ethanol. At very high concentrations—about 62 percent—the chemical is toxic and can kill upon contact. When the amount is lowered to less than 1 percent, bacteria tend to benefit as ethanol helps boost their energy production. But in between these two extremes, the reaction varies among species. Some may be tolerant of the molecule and do nothing. Others find alcohol a hostile presence and either move away from the area or die off.

A third group gets upset in the presence of alcohol and gears up for a fight, with consequences that can be disastrous for health. When a fighting bacterium recognizes the presence of ethanol and senses the concentration is too high to be useful, it produces several toxins in the hopes of killing off whatever happens to be producing the chemical. In the environment, the levels of the offending molecule can be quickly reduced. In humans, however, the process is unending because the source of the problem is not the gut but the brain. As long as the person keeps drinking, the bacteria will continue to pump out the toxins in the hope of ending the scourge.

In the meantime, the cells being attacked by the toxins alert the immune system to a growing problem. For some people, this can lead to gas and diarrhea and even vomiting. But if the binge continues, the reaction will go systemic and force the entire body into a state of illness. The body becomes feverish; the face, eyes, and hands become red; and the brain prepares for an oncoming battle. As anyone with a cold or flu knows, the effects are changes in mood, sleep disturbances, and even loss of appetite and social interest.

Every once in a while an overindulgence in alcohol offers no major concern. But regular bingeing can lead to long-lasting illness. And there's another problem: with each binge comes a change in microbial diversity. Many of the good bacteria we need to keep our bodies in balance disappear, while those that choose to fight thrive. The imbalance impairs the metabolic, cardiovascular, and even psychological systems by forcing them to stop normal activity and assume a defensive posture. This in turn can lead over time to chronic diseases, including liver failure, diabetes, and even depression—all classic symptoms of alcoholism.

But there is one more vicious consequence: over time, the immune system will become used to these toxins and learn to expect them. These molecular cravings grow into physical ones and increase the need to drink. If alcohol isn't consumed, the depression worsens, as does the likelihood of other symptoms, including anxiety and nervousness. All of these are considered part of the withdrawal process and can be almost impossible to control.

Fortunately, these effects are not permanent. The time required for the immune system to return to normal is only about three weeks. Cravings can be managed psychologically, with counselling, and microbiologically, through the restoration of microbial diversity. The immune system will stay balanced as it heals, and the mind will find itself less dependent on alcohol and keener on living a healthier life.

MICROBIAL MOOD KILLERS A fighting bacterium should never be dismissed. Though small in stature, it has an effective attacking strategy and can really cause damage to any unsuspecting cell. The trouble can become systemic as one fighting bacterium multiplies into millions or even billions.

The primary bacterial weapons are toxins, and they have an entire arsenal at their command. Some focus on killing host cells, but others initiate a cold war with the entire immune system. They deliver chemicals known to antagonize our human defences and send the immune system on a path towards indiscriminate destruction. This option is probably the worst for us, as it forces inflammation.

Inflammation can be a good thing, as we've seen, because it quickly identifies any trouble and sends troops to the location to resolve it. But when there is no actual invasion, inflammation can become chronic—essentially, the system waits for a non-existent attack—and spread to other parts of the body, including the brain. When this neuroinflammation occurs, it brings about several changes in the way we act psychologically, including lack of appetite, changes in motor function, lethargy, and a negative mood.

The worst effect of neuroinflammation is its inability to control itself should a threat never materialize. Until the bacteria causing the trouble are either eliminated or reduced to a safe number, the defensive action will go on without end. The situation can get worse, leading to depression. Medications may help, but they are only Band-Aids. Pharmaceutical antimicrobials may also help, but they also kill off friendly bacteria, particularly in the gut. The best course, then, is to restore diversity in the gut so the signals from the factions causing the inflammation fade over time. Whether you use diet, continual adherence to pre- and probiotics or simply adopt a healthier way of life, the goal can be achieved.

But a word of caution. It takes time to heal, and just one high-fat, sugary meal can annoy the bacteria in our guts and trigger inflammation. The same has been seen with binge drinking. Inflammation can worsen the effects of a hangover. In addition to the painful effects of dehydration, it brings on symptoms normally associated with a cold or the flu.

If a person starts to eat fatty foods regularly or develops a drinking habit, the consequences can be serious. Inflammation may become the norm, and that can eventually lead to depression.

NATURAL HIGH If they aren't asking me how they can live forever, people usually want the answer to another common question: "How can I find microbial happiness?" Until a few years ago, my answer was simply this: "Keep the good germs close and kill the bad ones." I know, I know—not really helpful. Today, I have another answer: take psychobiotics. This may at first sound like science fiction, but research has shown certain probiotic species can affect our mind, improve our mood, and calm us in moments of stress and anxiety.

Psychobiotics are similar to antidepressants in that they can control the level of chemicals recognized by the brain. But they are quite unlike their pharmaceutical counterparts, which function to artificially maintain the levels of these happy chemicals. Instead, these bacteria keep those levels high naturally. by actually producing many of these good-mood molecules as part of their day-to-day activities in the gut. The excesses are offered to the gut, which readily take them in. Once absorbed, the molecules trigger a cascade of neurological signals from the gut to the brain resulting in emotional balance and, quite possibly, bliss.

Let's say you're looking for mental balance; you need serotonin. It's known as the mood drug and needs to be in high enough concentrations to keep a person emotionally balanced. When we are depressed, we tend to have lower levels of this particular chemical, leading to a range of problems including reduced appetite, lack of memory, and a tendency to want to be alone. Many antidepressants block the body from using serotonin, thereby ensuring it stays at a decent level. But psychobiotics can produce the stuff or signal the body to do so. This can generate more than enough serotonin to keep the body balanced.

If you're suffering from anxiety, you may want to try a gamma-aminobutyric acid, or GABA. People who feel anxious or nervous tend to have low levels of this chemical in their brains, and as a result they cannot control the psychological turmoil. The chemical itself is already sold in pharmacies to control nervousness and keep a person calm. The bacteria form this chemical inside their cells to produce energy and keep those cells alive. When the nutrient supply drops and the cells begin to rest, the GABA is no longer needed and is sent out of the cell. This can mean a surge in concentration just waiting to be grabbed up by the intestines and sent to the brain.

If your goal isn't to calm down but get high, several bacteria have the uncanny ability to produce molecules that act in the same way as marijuana. The marijuana plant achieves this with a chemical known as tetrahydrocannabinol (THC). THC interacts with the cells of the body to produce the chemical responsible for the high, dopamine. When this gets to the brain, it triggers that feeling of relaxation and calm, as well as the urge to giggle and eat junk food.

The bacteria, on the other hand, act in a different way. They first produce and release dopamine in the gut. The effect of that is an increase in the absorption of the drug and a natural and potentially sustained high. Granted, it won't come near what THC can do, but that's actually a good thing. It means that a lower dose of the drug is regularly provided, giving a more sustained sense of happiness. But more importantly, it costs less than the weed and is perfectly legal.

It will be a while before we see psychobiotics on the drugstore shelf. They still have to go through various clinical trials and be approved by governments. But in light of the potential side effects of traditional psychiatric medications, the future for a more natural high appears to be bright. Psychobiotics may one day offer a daily dose of sunshine without a prescription.

ADDRESSING ALZHEIMER'S Microbes have been found guilty of contributing to the disastrous effects of Alzheimer's. Their mechanism isn't about direct infection, mind you. Although one particular form of herpes has been shown to cause Alzheimer's, the microbial connection is usually indirect, relying on a natural stage in the progression of the disease to cause troubles.

One of the steps recognized in the development of Alzheimer's is the breakdown in what is known as the blood–brain barrier. It's a mechanism we all have that prevents both microbes and certain chemicals from entering the brain and causing damage. As we get older, this barrier tends to weaken as a consequence of age.

Without this important control in place, the brain can be infiltrated by a wide variety of toxins. Once they get into the brain itself, they can kill the very cells we need for memory and normal bodily functions. As the killing continues, the brain can develop spaces, known as plaques, which can then grow and destroy even more of those essential cells.

This is where microbes play their part. Foes in the gut can produce a chemical known to be involved in the formation of plaques called amyloid. These species can also signal other cells, both microbial and human, to do the same. Over time, as its concentration increases, amyloid may head for the brain, crossing the now weakened blood brain barrier and worsening progression of the disease.

When we're healthy, there's no cause for concern—our defences can deal with the amyloid and get rid of it before it causes any harm. But if the immune system is weak, this molecule can accumulate in the body and may possibly enter the brain, making the situation worse.

Even if the molecule doesn't head to this sacred space, there may still be trouble. Because of the need to destroy amyloids, the defences go into overdrive. Unfortunately, this means inflammatory damage. Should this occur in the brain, the effects only worsen the situation and quicken the progression of the disease.

There is, however, a chance that microbes can offer some good in the prevention and slowing of Alzheimer's. The key is to focus on the good bacteria, our friends, rather than on those intending us harm. Once again, it seems that having a balanced microbial population in the gut may help keep us safe from this disease.

The first step is the formation of chemicals called polyphenols. Quercetin, resveratrol, and rutin have all shown benefit and can be found in many foods and drinks, such as fruit and even wine. We can also get them through supplements on store shelves. The trouble is that our bodies are not particularly good at absorbing these chemicals, and most are simply lost in feces.

But good bacteria—some of which are probiotics—can help in this process by increasing the levels of polyphenols found in the blood. Some of these may make it to the brain to counter the formation of the toxic amyloid.

The other way these good bacteria can help is by preventing the immune system from going out of control in the event amyloid is discovered. This action may not prevent the disease, but a more balanced attack can help to reduce the damage to the cells in the brain. This process is important for everyone but especially for the elderly, as they tend to have less good bacteria in their intestines than do younger people.

A note of caution: although probiotics can offer some assistance, they require daily supplementation in the form of fibres and polyphenol-rich foods. These should be ingested every day to keep the gut microbial population balanced.

YOU NEVER DIE ALONE The microbes that coexist with us stick around long after our existence is at an end. Even if you are no longer pumping blood, breathing oxygen, or sending electrical signals, you still present the perfect home for bacteria. The only difference is that you also become their food.

It's a natural shift in the balance between bugs and the body. When we are alive, we are able to keep microbes contained within certain regions, like the gastrointestinal tract and the sinuses. Our various tracts form tight cell-to-cell barriers to keep bacteria from entering unauthorized zones. Areas exposed to the environment, such as our eyes, lungs, and guts, produce mucus to trap the bacteria and hold them in place. On the inside, the immune system protects the organs by examining all microbial visitors and killing those not considered to be friendly. Altogether, these systems maintain a bacterial balance so we can live in harmony.

But as soon as we die, that balance is lost. The microbes are now able to travel at will, and they have a rich supply of nutrients. Everything is potentially food and the microbial population begins to flourish. In the meantime, our bodies initiate a process of self-degradation, in which our own digestive enzymes start to break down our tissues, making them easier for the bacteria to digest.

As this happens, our bodies change. The first stage is bloating and the production of gas (the latter turns our skin colour a greenish-black). When we're alive, microbial volatiles are ejected as flatulence. Deep inside the tissues, however, there is no such exit route. The second is loss of tissue as it is eaten away. It's akin to flesh-eating disease. Within five to ten days, the organs and blood begin to liquefy as a result of microbial enzymes (the same ones we rely on in life to break down proteins). Over a few weeks, the skin finally breaks down and, eventually, disappears.

Most of the microbes conducting this breakdown are the ones we consider our friends over the course of life—the ones that help us digest our food, keep our skin safe, and balance our immune system.

In essence, decomposition is the final task of our friends. They helped us during life, and in death they help us to be, in a way, environmentally sustainable.

Clostridium botulinum

7. FOOD

UNDERSTANDING UMAMI Of the five basic tastes—salty, sweet, sour, bitter, and umami—it's the last on the list that's hardest to pinpoint. Scientifically identified in 1908, umami is Japanese for "pleasant, savoury taste." It's vague, I know. But once you've tasted it, you know it.

Chemically, that unique flavour is the result of only a handful of molecules, principally glutamate. That's why chefs love to use monosodium glutamate, better known as MSG. When added to a sauce or other food, this salt gives depth to the basic flavour. The resulting sodium content is a health concern, of course.

Thankfully, MSG isn't the only source of umami. Naturally umami-tasting foods are everywhere, and include truffles, tomatoes, green tea, eggs, shellfish, and seaweed. Glutamate and a group of similarly flavoured chemicals known as ribonucleotides can also be made in the lab and added to food products. And there is another way to add the taste, without adding chemicals. It's the ancient process known as fermentation, and it offers significant health benefits.

Fermentation is the deliberate spoiling of food with known species of bacteria or fungi in order to prevent rotting from more harmful microbes. Before the advent of refrigeration, fresh foods had to be preserved to make them last. Curing with salt and pickling in vinegar were two methods of preservation, but these both radically altered the taste. Fermentation had the two-pronged advantage of preserving a food's flavour while giving it a little umami enhancement.

The enduring popularity of fermented foods since antiquity has to do with ease of preparation. The simplest method has only a few steps.

1. Chop, dice, or grate the food source—let's say cabbage for sauerkraut—and then place it into a container.

2. Add enough water to immerse the mixture. Add salt, but not too much—this isn't curing, after all.

3. Use a porous cover or lid, such as cheesecloth, and add a weight such as a rock to keep everything under water (this helps to prevent any fungal growth from the outside air).

4. Wait.

The entire process can take anywhere from a few weeks to months, depending on the desired strength of flavour. When ready, take off the weight, remove the cover, and enjoy that umami taste.

Fermentation is a fantastic way to improve food safety and security, as it acts as a means of preservation. Granted, it is "spoiled," but it still imparts a pleasant, albeit sometimes acquired, taste. But there is an even better advantage to having these foods in our diet: they are some of the healthiest foods, particularly when what is fermented is raw, providing a wealth of benefits. When we eat something fermented, particularly if the food is raw, we take in a variety of healthy by-products.

Analyses of several fermented food products—including dairy products, grains, vegetables, meats, and even fish (yes, fermented fish!)—reveal that many of them contain antioxidants, anti-inflammatories, and even metabolism-regulating molecules. Fermented food is so good for us not so much because of the nature of the raw product but the fermenting organisms. The bacterial and fungal species are either the same as or related to the healthy bacteria we already have in our gastrointestinal tract.

Although this may appear to be an odd coincidence, from an anthropological perspective, it makes perfect sense. These bacteria are needed for our overall health, but it's not an easy task to keep them in high levels in the gut. The intestines are an ecosystem and very difficult to control on a regular basis because of factors ranging from poor diet to infection and antibiotic treatment. Fermentation, on the other hand, is a more controlled process in which we can cultivate these microbes, much like a microbial farmer, and then introduce them to the body whenever we have the chance. The levels of good bacteria can remain high enough to impact health even if our bodies are not up to the job. This is particularly helpful in those with weakened bodily functions, such as the very old and the very young.

The benefits of fermented foods may explain why our tongues have devoted one-fifth of their capacity to these food types. Their importance to health is without question, and they should be included in everyone's daily diet. Fermentation can be done at home, and the number of naturally fermented items available in grocery stores continues to grow. When purchasing a fermented product, make sure the ingredients include a bacterial starter culture (some may even list the Latin names). And if it's possible, get the foods raw; pasteurized varieties will have the same chemical benefit, but the good bugs will be gone.

THE FERMENTED THREE The shock of accidentally taking a gulp of spoiled milk is hard to overstate. Instead of being greeted by the expected sweetness and smoothness, the taste buds are attacked by a sour, acrid concoction usually accompanied by those revolting chunks. Although the milk is safe to swallow, spitting it out is easier.

How does milk sour? When the liquid is exposed to air, a variety of bacterial and fungal species find their way in and thrive. As this happens, the wonderful taste of the milk changes as the sugars and fats are used up and turned into much less tasty by-products. The acidity increases, leading to a separation of solids from liquid. The entire solution becomes unpalatable and finds its way to the sewers.

There are a few ways to prevent this. The first is pasteurization. This involves heating milk to a temperature known to kill off most microbial life and then quickly cooling it to maintain the chemical composition. This is the standard method for preserving milk and is commonly used to lengthen the shelf life. The other way is to change the milk altogether by purposely spoiling it through fermentation. The end result is no longer milk but a product like kefir, yogurt, or cheese.

At their most basic level, these three dairy products are similar. They are fermented, they are made from dairy, and they taste good. But they differ in two important ways. The first is their water content. During the manufacturing process, kefir remains fluid (although it's thicker than milk itself). The water content of yogurt is significantly less as a result of partial solidifying during fermentation. And cheese requires a preliminary step of separating the solids from the liquids (the curds from the whey). The second difference is the makeup of the bacteria responsible for the fermentation. Each of the three requires a specific number of bacterial species to produce a combination of flavour, smell, and texture.

Although the microbial populations involved in making these three products may differ, we know that at least one species in the mix will be a lactic-acid bacterium. This microbe feeds on sugars such as lactose in milk, and in doing so forms a by-product called lactic acid. This chemical is a fantastic preservative because it has antimicrobial properties to prevent the growth of other species. This isn't the only benefit—these species also help to improve the overall nutritional value of milk products. They enrich various vitamins so they are more easily absorbed into the body. They produce a variety of fats and proteins our bodies need to survive. They also make a number of chemicals our immune system recognizes as signals of calm. Some of these bacteria can even help to reduce cholesterol levels by trapping the molecule and disposing of it in the feces.

It's no surprise that people keep jumping on the bandwagon for fermented dairy. Today, companies all over the world are making fermented dairy products to satisfy demand. The upscaled manufacturing process is complex and involves hundreds, if not thousands, of litres of milk each day, and must also be overseen by microbiologists. These trained individuals make starter cultures for each batch to ensure that the bacteria are the same every time. For the manufacturer, this is an important step, as several species are involved in milk fermentation and each provides a different combination of tastes and consistencies. To keep a product uniform, the same bacteria need to be used.

Fermented milk can still be made the old-fashioned way at home. The process is surprisingly easy. Just take a container of milk and add bacteria from a previously fermented batch or a starter culture. Keep the liquid covered to prevent any environmental bacteria or fungi from entering, and then wait. The milk will change as it becomes thicker and possibly more solid. The taste will also change, but if the fermentation is done right, the milk will have more depth. The time required will differ depending on the bacteria and the expected quality of the end product, but the simplest fermented-milk product, kefir, takes only about twelve hours to make.

BITTER SWEET I have a sweet tooth and am proud of it. Sweetness is, after all, important for survival—and one of only five tastes we can detect on the tongue.

There was an evolutionary need for us to detect sugar quickly. For thousands of years—before the advent of sugar production—finding this nutrient was difficult. We required a quick and rapid way to identify it. Our taste buds gave us the ability to recognize sugar instantly (and eat it very soon after).

But not all sugars are alike, and only one is really important for our health: glucose. It's a simple sugar, made up of one molecule that has a sweet taste and packs quite a bit of energy. As soon as glucose molecules reach the small intestine, our bodies set in motion a protocol to pick up as many of them as possible. Once a single molecule is grabbed, it's sent to the blood, where it can nourish our human cells.

Our bodies can't pick up all the glucose molecules in the intestines, but that's not really an issue, because the bacteria living in the small intestines are more than happy to use whatever glucose remains. They too quite enjoy sugar and are also quick to take it in for their own energy. If there's any left, it will head to the colon, where other species will use up what they can. If everyone plays their part, the ingested glucose molecules will be mostly used up, with only traces left behind to be excreted in the feces.

This is the perfect scenario, but it's rarely seen in you and me. Instead, shifts in balance occur regularly and change how our bodies and bugs react. The most common issue is overconsumption of sugar. Our bodies can absorb only about ten grams of sugar each time we eat. But most sweet foods have far more than that, so our bodies become overloaded and cannot keep up. The bacteria in the small intestine might be able to help with the excess, but this too can be trouble. Friends will do their part, ensuring that no toxins are made. But foes will not provide us with the same courtesy. As they take in the glucose, they will begin to overgrow and might start to look for places to invade. They could even begin to attack the friends, as well as the intestinal wall. Thankfully, the intestines are used to this trouble and can fend off an attack. But the friendly bacteria may not be so fortunate, and their numbers could drop.

When the excess glucose makes it to the colon, the bacteria will happily take up close to 85 percent. But the way these bacteria use the sugar means they take only about 10 percent of the energy available from each molecule. The rest ends up being sent out of the cell and, more often than not, into us. Although it does nothing for us in terms of satisfying hunger, it does end up increasing the number of calories we get from the food.

How much? Well, if you take in too much sugar, you could end up with an excess from the colon of as much as fifty calories a day. Those calories add up: 1,500 in a month, 18,250 in a year. Now consider that every 3,600 calories equals a pound of fat and you have a recipe for weight gain.

Fructose is another simple sugar most of us encounter every day. The problem is, our bodies are not a fan of this chemical.

It's sweet, yes, but it has almost no nutritional value. The intestines scoop it up and send it straight to the liver for processing. This detour is needed to make the fructose useable to the cells. But the process is inefficient and leaves plenty behind for the bacteria.

If only a few grams of fructose are taken with each meal, there's really no issue as the body and the bugs will be sated in a balanced way. But if the amount goes up to around twenty-five to thirty grams per serving, there could be a problem. The bacteria will be flooded with the fructose molecule, but only certain species can actually use it. The rest are out of luck. This could end up in an imbalance in favour of the fructose users. It may come as no surprise that these bacteria also tend to provide more calories to us, and thus can increase the chances for weight gain.

There's an easy way to avoid microbial troubles with glucose and fructose: just don't eat too much of them. Biologically, physiologically, or microbially—too much sugar is trouble in every way.

ARTIFICIAL FATTENERS Artificial sweeteners can be found pretty much anywhere in the grocery aisles, usually as an ingredient of a product with "diet" in its name. But the sweet dreams of easy weight loss can quickly sour when you consider the drawbacks. These sweeteners carry an increased risk for both high blood pressure and diabetes. And they can actually lead to weight *gain*.

Sugar is the fastest way for us to obtain energy. It is necessary for proper metabolism, and it is the most important nutrient for our cells. When we cut sugar from our diet, our body must find ways to deal with its absence. One of the most effective is to store energy in the form of fat. As the sugar starvation continues, the fat increases, leading to additional weight.

There's more to the picture. Hunger is controlled by a series of hormones, the most important of which is insulin. It makes us feel hungry for glucose. If we choose to ignore this urge by using artificial sweeteners, hormonal levels will rise to compensate for the continuing lack of the sugar. As this happens, the cells of the body stop responding to these involuntary hunger pangs; they refuse to listen to the insulin and will turn away from glucose even when it is present. This condition is called insulin resistance, and it's the first step in the development of Type 2 diabetes. Although small bouts of resistance can be reversed, this can become a permanent condition requiring lifelong medical treatment.

If this isn't bad enough, there happens to be a third consequence: many species of bacteria react badly to the presence of artificial sweeteners. They need a sugar such as glucose to survive; without it, they end up starving to death. Whole populations of bacteria can be destroyed over time. In their absence, other species take up the room and continue to thrive as if nothing happened.

This change in diversity may appear at first to have no effect on the human body, but a closer look at the bacteria taking over the gut reveals they are not our friends. Instead, they are self-centred, with little care for their human counterparts. They are not officially pathogenic—they don't cause illness—but they do tend to have a beef with the immune system and pester it continually. The immune system has no choice but to stand at the ready for an attack, even though none will come. This leads to inflammation, and as we've seen, inflammation will over time spread to the rest of the body and force the various systems to stand at the ready too.

This causes all sorts of problems, from the simple (more frequent pain in certain areas of the gut and an increase in diarrhea) to the complex (changes in mood, a lack of mental clarity, and a possible worsening of the disrupted sugar-and-insulin balance). But perhaps the most vicious problem has to do with taste. With these bacteria in place, the ability to taste food begins to dampen, and certain flavours may become bland or unwelcome. Sufferers may require even higher doses of chemicals such as salts—and yes, more artificial sweeteners—to make up for the problem. Over time, the only way to sate an appetite is to eat more chemicals.

There are safer alternatives. Naturally sourced sweeteners are increasing in number and may one day take over from their artificial counterparts. One of the most effective and beneficial is a sweetener made from the stevia plant. The active flavour comes from an assortment of molecules known as steviols. They are not sugars, but they do contain several sugar-like elements, giving the tongue the sweetness it needs. Once the chemical gets to the gut, it interacts not only with the bacteria but also with human cells. They are told to keep calm and not to ready themselves for an attack. As for the bacteria, they can break down the steviols to use as food. This means that none of the friendly species will end up starving, and those self-centred gluttons are kept in line thanks to the stable diversity inside the gut.

HEAVENLY HONEY Of all the foods made by nature, one of the most beneficial is honey. The thick, sweet liquid is mainly composed of fructose and glucose, as well as a chain of the two molecules known as sucrose. But there's so much more to honey, including proteins, antioxidants, and minerals. Perhaps most importantly, it is fermented, meaning it contains several bacteria (when raw) and their by-products. Fermentation is an important step in preserving the liquid as it is one of only a handful of natural foods that do not go bad.

Honey is made from the nectar of various plants. Bees collect the liquid as they travel from flower to flower and store it in a specialized stomach. Within this organ are a number of bacterial species that are fed by the nutrients in the collected nectar. When a bee returns to its hive, it regurgitates the microbe-rich solution into specialized compartments. Here, the honey matures as the bacteria use the sugars and other nutrients for growth. Proteins are broken down into more usable amino acids, the sugars are transformed into tangy acids, and the minerals are modified so they can be more quickly absorbed.

There is another component: natural chemical preservatives. The bacteria within the maturing honey produce several antimicrobials known to be toxic to pathogens. Many of these are commonly found in household disinfectants, such as hydrogen peroxide and formic acid. The concentrations are so low they pose no threat to humans or bees, but there is enough to keep any visiting invader from growing.

The matured liquid is the perfect way to sate any worker bee and is readily taken in after a hard day out in the field. But the bees are not the only benefactors of this wondrous substance. We humans have a love for the stuff as well, as it not only naturally satisfies a sweet tooth but also helps keep our bacterial population diverse and happy.

As soon as honey enters the mouth, it spreads out, covering the cheeks, the gums, and the teeth. The antimicrobials contained within can attack bacteria known to cause cavities and improve the overall microbial diversity. The presence of other freely available nutrients, including calcium, phosphorus, and even fluoride, helps to keep the teeth strong and resistant to acid breakdown.

After it's swallowed, the honey coats the stomach, balancing out acidity. This can prevent damage to the stomach wall during acid production. If any problems have occurred, the liquid can stimulate wound healing to remediate the condition more rapidly. Regular use of honey can also prevent heartburn, as various antioxidants block an overproduction of acid.

When the liquid reaches the intestines, the body begins to absorb the glucose and fructose, giving a quick energy boost. But unlike sodas and candies, which are also rich in these sugars, honey gives no sugar high, and that means no crash. Those longer chains of sugars require more time to break down, which ensures that the body is continually provided with a balanced level of sugar. The slow and thorough process of digestion also helps to balance appetite and cravings so we don't feel the urge to eat as often.

As for the bacteria in the gut, only those able to survive the antimicrobials can join the party. There are several species capable of not only resisting but also thriving in the presence of honey. Thankfully for us, these are the same ones known to provide us with optimal health benefits. In combination with honey, these bacteria can improve overall health by reducing the risk for a variety of chronic ailments associated with sugar. These include weight gain, obesity, diabetes, and impaired liver function.

Getting honey into our bodies is not all that difficult as it can be found in almost every grocery store. But not all honeys are the same. Many are pasteurized to kill off the bacteria they contain. Though pasteurization is excellent for food safety, to reduce the chance of contamination, it lessens the natural benefits of honey. The best option is to seek out raw honey so that all the nutritional and microbiological benefits can be gained.

HOW SWEET Xylitol is a rising star in the world of nutrition and health even though it has been around for over 30 years. This safe, natural chemical looks like a sugar molecule but doesn't have any of the calories. It's extracted from hardwood trees, purified, and used as a sweetener, giving us all the joy of sweetness without any worry for weight gain.

Xylitol is also good for our microbial population, helping to keep it in balance. The chemical is similar to fibre in this sense. Many friendly species can easily take it up. As for foes, they are forced to starve, meaning their population will quickly decrease.

But starvation isn't the only way to eliminate bad bugs. Xylitol also breaks down the outside wall structure of certain pathogenic bacteria. Just a few hours of exposure can be devastating as the chemical eats away at the microbial barriers. Continued exposure soon kills these pathogens and restores balance in the gut.

Because of the effects of this marvelous chemical, you probably will see it appear in the ingredient lists of foods and drinks known to be naturally sweetened. You'll also notice it in chewing gums. The chemical is known to reduce the number of bacteria responsible for harmful plaque. Although it is no substitute for good brushing and flossing, it is a great way to help keep those teeth and gums healthy.

THE DIVINE BENEFITS OF CHOCOLATE My favourite food of all time is chocolate. I've even gone to the Amazon jungle to learn how the seeds of the cacao tree are turned into this miraculous substance. It's amazing to sample each step, and to learn exactly how chocolate gets its taste.

In some parts of the world chocolate is called the food of the gods. If you ask me, though, a more appropriate description would be "food from the microbes." After all, before any of us can enjoy chocolate's unique combination of sweet and bitter flavours, bacteria must work hard to magnify those flavours and boost the nutritional benefits to us and our microbial inhabitants.

The chocolate-making process has been around for centuries. The seeds of the cacao plant must first be liberated from large pods about the size of two fists. The pods contain not only the seeds themselves but also a rich, whitish pulp. Like other fruit, this pulp is rich in nutrients and can be harvested for use in beverages and powders. But for the chocolate producer, the real value lies in the pulp's ability to grow fermenting bacteria. Without them, the quality of the chocolate inevitably suffers.

Over the next seven days, the seeds are left in the pulp and allowed to dry, usually in the sun. During this period, the pulp grows a rich diversity of bacteria, yeasts, and fungi. As their numbers grow, they begin to attack the seeds, killing them and initiating a form of decomposition. A number of chemicals are produced in the process. Preservatives such as lactic, citric, and acetic acids help to prevent pathogen growth. Organic compounds, including caffeine and the aptly named theobromine (the chocolate plant's Latin name is *Theobroma cacao*), provide both bitterness and a greater depth of taste. A number of antioxidants and anti-inflammatories are also produced, as are sugars (although they are mainly fibrous in nature).

After fermentation is complete, the beans are roasted and then ground to make a brown chocolate liqueur. Although some of the chemicals made by the bacteria will disappear, most are heat-stable and will make it to our mouths. The final product may contain over five hundred different chemicals contributing to a variety of flavours such as fruity, floral and astringent.

The best form of chocolate for human consumption contains 70 percent of the liqueur with the rest comprised of low amounts of sugar and fats. The senses are maximized as the aromas fill the mind with pleasure and the taste provides a complex yet enjoyable experience. But while those few seconds may be heavenly, the benefits don't end there.

After they've been swallowed, those five hundred or more chemicals get to work. The antioxidants and anti-inflammatories help to keep the immune system in the gastrointestinal tract calm. The organic molecules will stimulate the metabolic system, and even spark an energy surge (thanks to caffeine). Some chemicals, such as flavonoids, are also known to affect appetite and with continued ingestion may improve cardiovascular health and even calm anxiety.

The bacteria in our intestines also feel the effects of chocolate. The fibre contained within is the perfect food for a number of species, and they readily eat up these molecules. But that's not all they like. Several of the organic and aromatic compounds may still be fermented even further, and so the bacteria go to work on these. As they do, a greater variety of antioxidant and anti-inflammatory molecules are produced. Many of these end up heading into the bloodstream, where they can improve sugar balance so as to prevent highs and crashes.

With all the benefits chocolate offers, there is little wonder it is so ubiquitous in our grocery stores. But not all chocolate products are the same, and most have little to no value. It comes down to the actual amount of cacao they contain.

The best chocolate is 100 percent cacao, though it is rather bitter. Taking the concentration down to 70 percent provides enough room for sugar and fat (although these too should be naturally sourced, such as raw cane sugar and polyunsaturated fats). If, however, the value goes down any lower, the potential for positive returns drops. The worst possible option in terms of health is white chocolate; it has absolutely no cacao whatsoever.

FRUIT VS. JUICE Which is better for you: juice or whole fruit? The debate has been going on for ages, and there seems to be no end in sight.

Those who prefer whole fruit claim it's the only way to get all the nutrients. Those who prefer juice argue it combines the benefits of whole fruit with greater convenience. From a purely microbial perspective, whole fruits win hands down. Not only are the bacteria getting all the benefits of the juice, but they are also getting fibre as well as a selection of phytochemicals and antimicrobials. This allows for better diversity and a chance for friendly bacteria to thrive.

There are also other issues with juice. First, there is usually little to no fibre content; only simple sugars are available. While all bacterial species will enjoy the sugar rush, our foes tend to have a better time getting to these molecules and using them up. Their numbers rise, while the number of friendly bacteria may drop. Also, fruit juices have to go through an extra processing step, and they may pick up a few harmful bacteria along the way. Introducing these bad bacteria into the gut can cause a host of problems, including gastrointestinal infection.

Still, when I'm rushed I reach for a glass rather than a piece of fruit. But my glass contains a smoothie. This delicious blend of puréed fruit is easy to make, quick to take, and it provides all the goodness of whole fruit. It's a delicious way to start the day—and personally, I have another smoothie just before bed. The nutrients and fibre keep the bacteria happy while I head off to sleep.

As for which fruits are best, there's no need to compare apples to oranges. Almost any fruit—or nut—will do. But bananas, apples, coconut, strawberries, blueberries, cranberries, and almonds are all chock full of nutrients for both you and your bacteria. You can usually find several varieties of smoothies in grocery stores—just make sure the primary ingredients are not sugar or water. Or just get a blender and make them at home. That way, you can add some yummy probiotic yogurt or kefir to the mix and get twice the goodness.

HOW NOT TO GO TO SEED When I was growing up, I had no idea what a pomegranate was. But over the past few decades the fruit seems to have taken over as the number-one health food. It may once have been dismissed as a fad, but today it's clear that the red orb with all those seeds is here to stay.

The key to the pomegranate's many benefits, which include cancer prevention, better bone health, lower blood pressure, and healthier teeth, is its large supply of polyphenols. These chemicals, which are found in almost all fruit, help our bodies remain balanced thanks to their anti-inflammatory and antioxidant properties.

One particularly beneficial polyphenol is called ellagitannin, which was originally found in gall nuts but eventually in pomegranates. It is simply loved by our gut-friendly bacteria, which use it for metabolism. Yet many unfriendly species, including pathogens, react to ellagitannin as if it were a disinfectant like bleach. They simply cannot survive in its presence and end up dying off. This type of selective killing action is a great way to help maintain the balance of bacteria in the gut.

The best way to get these ellagitannins into you is to eat the raw fruit. If you feel this isn't the easiest task, what with those hundreds of seeds, go for an extract instead (though preferably not one from concentrate). It offers many of the benefits of the real thing.

STARCH CONTRAST Have you noticed potatoes are wonderful when cooked but absolutely horrid when raw? The same goes for corn. Cook it up and it's amazing, but straight off the stalk you may prefer to go without. It comes down to the sugars in the vegetables in the raw state. Instead of being simple, they are in the form of unsavoury starch.

Chemically, starches are classified as sugars, but they're not like simple sugars such as glucose and fructose. They are instead long chains of glucose held together by strong bonds that can resist heat, making them perfect for cooking. They can be found in a number of vegetables and legumes, such as the aforementioned potatoes and corn, as well as beans, root vegetables, and lentils.

In its natural state, starch is not a nutrient for us. It has to be broken down to glucose for energy to be harvested. Our bodies are used to this and produce a specific molecule, an enzyme called amylase, to break those bonds. Amylase is primarily found in saliva and is used—in combination with chewing—to break those starch chains. In the intestines, other enzymes help with digestion, but their contribution is minimal. The rest of the work is accomplished by bacteria.

Microbial digestion of starch follows a similar path. The bacteria form a different kind of amylase that is far more active than the type humans produce in the saliva and intestines. The enzyme can quickly break those bonds, freeing up glucose for use. Whether the process happens inside or outside of the cell, the effect is the same—the local glucose concentration rises, and the bacteria live and thrive.

In the environment, amylase is quite common among bacteria, but in the human gastrointestinal tract, there are only a few species capable of making the enzyme. Only a handful of the five hundred to a thousand bacterial varieties found in the intestines are able to use starches. The rest of the bacteria—as well as intestinal cells eager to be fed—have to wait for the enzyme to do its job before they can benefit. Unfortunately for them, there isn't much left behind; as soon as the glucose is formed, the enzyme-producers grab up the molecules for their own use.

For us, this is a good thing as the process slows down digestion. The lowered availability of glucose means the cells cannot gorge themselves on sugar, leading to spikes in the blood. The more measured approach means we can go longer without feeling hungry. This can help lower the risk of weight gain caused by an excess of energy. The longer transit time also helps to improve absorption of nutrients, including calcium, magnesium, zinc, and iron.

Starch in the diet brings another long-term benefit, thanks to its resistance to digestion. To process starch, the intestines are forced to produce more digestive enzymes and bile. If unused, these chemicals are returned to the liver. But in the presence of starch-loving bacteria, fewer chemicals make the round trip.

As this happens, the liver senses a need for bile and commands the body to produce more, using the only raw ingredient available: cholesterol. Over time, the level of cholesterol goes down, as does the amount of lipids in the blood. This can help to reduce the risk of cardiovascular problems, including atherosclerosis.

But even with the amylase-containing bacteria, the actual benefits of starch ingestion are far less than 100 percent. As little as half of the actual nutrient value in starches is used by our bodies. The remaining nutrients are simply lost as they move down the gastrointestinal tract to the inevitable release in fecal matter.

Starch is found in most baked goods, as well as in raw vegetables and legumes. It's also in many flours and cereals. But because the flavour of starch is quite unpleasant, it's best consumed as one ingredient in a more complex and ultimately tastier dish.

THE K FACTOR Vitamin K is responsible primarily for blood coagulation (the K is for the German word *Koagulationsvitamin*), but it also has numerous other functions, including helping in the formation of proteins. Vitamin K is an essential part of life, but it's also one of the chemicals our bodies cannot make. The richest sources of vitamin K are plants like spinach and kale. But good amounts are also found in meats and eggs. Fermented foods also contain the vitamin, thanks to bacteria.

Many of the friendly bacteria living inside us make vitamin K. They do this as they digest fibre. As the molecule is prepared, our intestines happily absorb them and circulate them throughout the body.

The best way to get all the vitamin K you need is to ensure that vegetables (including fermented foods) are a part of your daily diet. Then there are those probiotic fermenting species. By having them in the gut, whether in the short term or as full-time residents, we can ensure that we're getting vitamin K we need. Mind you, the amount we take in from bacteria isn't enough to keep us going over the long term. But when vegetables are not available, the microbial contribution will definitely help.

FIBRE OF OUR BEING We are constantly being instructed by dieticians, nutritionists, and doctors to eat more fibre, a substance our bodies cannot digest. The molecule is a long chain of sugars held together by bonds the digestive system cannot break, no matter how hard it tries. This is a good thing.

The most basic use for fibre is bulking up the digestive tract. This helps slow down the digestive process, enabling our bodies to absorb more nutrients from other foods. The sheer size of fibre also maintains the solidity of bowel movements, reducing the frequency of loose and watery stools.

There is, of course, much more behind the benefits of fibre. The rest of the goodness lies with the bacteria living in the colon. They simply love fibre and use it as a fantastic nutritional source to keep them thriving. These species also have the ability to thank us for feeding them.

When a bacterium comes into contact with a fibre molecule, it releases a series of enzymes capable of breaking those strong bonds. Then the individual sugars are absorbed for a different type of digestion called fermentation. This process differs from regular sugar metabolism in several ways, but the most important for health concerns its waste products. When human cells use sugar, the end products are water and carbon dioxide. When these bacteria have done their work, the end result is a form of fat called a short-chain fatty acid. These are highly aromatic compounds giving off many of the smells associated with feces.

For human cells, which have no nose, these molecules are no problem. They regard these short-chain fatty acids as markers of good health. The most important of these molecular benefactors is known as butyrate. It's been studied extensively in the lab and in clinical trials to assess how it helps us. It turns out that butyrate makes a variety of contributions to our well-being, including maintaining the barrier function of the intestinal wall. When cells come into contact with the molecule, they are prompted to produce more mucus, which is one of the primary barriers against infection. As the mucus increases, so does the tightness of the cells themselves. This helps to reduce the likelihood of any wayward pathogens or other foreign substances getting into the tissue.

Butyrate also helps to calm down the immune system. The molecule has the ability to interact with human cells in the colon to let them know there is no need for a defensive posture. These cells then send out an all-clear to the rest of the immune system. This anti-inflammatory action is local, but the signal is carried throughout the body and even into the brain.

Butyrate's most important role may be in preventing cancer. In the colon, stressed cells may end up transforming such that they begin to reproduce uncontrollably. In some individuals, this may lead to colon cancer. Butyrate can prevent this malignant process by getting into these stressed cells and shutting down the cancer-causing activity. Some cells will return to normal, while others will die off. Either way, the chances of a tumour being formed are greatly reduced.

There is a final benefit to butyrate and other short-chain fatty acids: they can help you feel full for hours after each meal. These molecules interact with intestinal cells, telling them the bacteria are doing fine and there is no need for any more food. This is especially evident in the evenings. If you eat fibre before bed, the butyrate concentrations will keep your body sated throughout the night. That way, you won't feel the need to get up for a midnight snack.

One problematic source of fibre, by the way, is corn. It's high in nutrients, but not even gut bacteria can break down those kernels. That's why after a good corn roast, the next day's fecal output features those little yellow dots.

GOOD AND BAD FAT   We need fats to live. They are important for our health and should make up about 40 percent of our daily nutrient intake. But not all fats are the same. Depending on which ones we choose, we can be healthy or suffer from a myriad of chronic ailments.

Fats are long chains of oil-like chemicals. When put together, they can form a very strong structure like a dense forest. They are also flexible and can sway much like trees. This makes them useful for cell structure, as they can keep the shape of a cell while still having the ability to shift shape when needed.

Fats also store energy. In fact, each molecule holds over twice as much energy as a carbohydrate such as glucose. When fat molecules are broken down by an enzyme, that energy is released for use in another biological process. It's incredibly important to have these molecules when our ability to take in food and drink is hampered, or when we are overexerting our bodies.

Bacteria need fat for the same reasons. They use the molecules in their outside membranes for structural integrity. They also find significant energy resources in fats, and several species have the ability to break down these molecules and harvest the energy for other processes. Then there are defensive mechanisms, in which fats are turned into toxins to kill enemies.

But bacteria act differently depending on the type of fat. Some molecules are toxic to certain pathogenic species. The most toxic are the polyunsaturated fatty acids, or PUFAs. Some of the best PUFAs—such as linoleic acid and the omega-3 fatty acids—are found in fish and flax oil. In the intestines, PUFAs can inhibit and even kill the bad bugs, limiting their growth. PUFAs also stimulate the production of bile in the gut. This only adds to the problem for these foes, as bile is itself a weak antimicrobial. With continued ingestion of these good oils, the numbers of bad bacteria will eventually drop.

Saturated fats have the opposite effect. These molecules are extremely resistant to degradation and heat, making them great for food manufacturing. They are also highly abundant in fatty meats and dairy products such as butter and cream. Because they are so hard to break down, saturated fats are usually left alone by the body. This results in a lower amount of bile in the gut, making digestion even harder. And there's another downside to having less bile: higher cholesterol levels. Bile is made from cholesterol, and a lower production means there's more left in the body. Higher levels can then contribute to cardiovascular disease.

The situation worsens when the bacteria are brought into the equation. Saturated fats tend to be used up by many unfriendly bacteria and they can thrive. If they overgrow, they will initiate a search for more places to colonize. This is bad news for the intestines, which the bad bacteria attack in order to break down the barrier preventing them from exploring. Without a response from the immune defences, the bacteria can open up the intestinal wall and continue their travels. Thankfully, the body is keen to defend itself and does so by becoming inflamed.

If this happens once in a while, there is no real concern. But if the body is continually loaded with high levels of saturated fats, the attacks on the intestinal wall will continue without end, and eventually the bacteria will accomplish their goal. They can then travel to places like the blood and liver, where they can cause even more damage and instigate a war.

Our bodies are not keen on war but will fight if needed. The trouble is, war takes a toll on energy reserves. To ensure there is enough energy to fight, cells not essential to the battle are restricted in their intake. The level of glucose in the blood goes up, as does the number of energy-storing cells, known as adipose tissue. These cells can be found all over the body, but they are usually stored close to the liver, in the lower torso. Our weight goes up and the adipose area grows in girth, causing us to ask, "Do I look fat?" (Based on personal experience, the best answer is "No.")

To avoid the war and the weight gain, it's best to limit the amount of saturated fat eaten. This is not easy, considering how much of it is used in processed foods. Choosing more natural sources will help, as will focusing on lean meats. But there doesn't seem to be a limit to how many PUFAs can be ingested. They are going to be good for us and our friendly bugs, and help to keep those foes in line.

THE SECRET OF SOY If you want to add some savour and flavour to a meal, reach for the soy sauce. This black liquid offers an exciting combination of tastes to boost any meal. And no two brands of soy sauce are quite the same.

Soy sauce is of course made from the soybean, along with wheat, water, and salt. This concoction is at first only salty; it also offers for the most part no health benefits. But then microbes are added—a combination of yeasts, fungi, and lactic-acid bacteria called koji. Once they get into the slurry, everything changes as the microbes use enzymes to break down the various proteins, starches, fibres, and sugars.

There are two stages of fermentation in the making of soy sauce. The first involves fungi, which break down the wheat and soy. This lasts for only about three days. The second takes up to twelve months and requires lactic-acid bacteria and yeast. During this time, these species use up sugars, starches, and proteins to produce several acidic and savoury by-products. The yeast also makes alcohol, which helps to increase the flavour (think about adding wine to a mouthful of cheese or meat). After the fermentation process is complete, the mixture is harvested and refined to separate the solids from the liquid. The liquid is then pasteurized and eventually bottled for use.

The resulting sauce is good for taste and good for us. It helps digestion by promoting enzyme production in the stomach and intestines, which ultimately improves nutrient absorption. And when the sauce hits the intestines, many of the by-products will seek out and destroy unhealthy species of bacteria. These antimicrobial compounds are produced during fermentation to keep out unwanted species; they remain once they are ingested. A host of food-borne pathogens can be kept at bay thanks to the sauce.

But wait! There's more. Several by-products such as antioxidants interact with the gut cells. Many help to keep immunity balanced while others have been associated with lower blood pressure, blood thickening, and preventing cancer. Although these benefits are not solely related to soy sauce—they are based on lifestyle and diet as well—it can be an added advantage.

There is one issue with soy sauce, however: the high salt content. Even the "lower-sodium" varieties should be used cautiously by anyone avoiding salt because of high blood pressure or other health concerns.

If you want to forgo soy sauce, there are fermented alternatives that offer similar benefits. Of them, miso, is the most popular in Asian cuisine. This paste has the consistency of peanut butter and a rather unique taste. It's a must for many soups and also can complement salads and condiments. Soy yogurts and fermented beverages are also rich in beneficial nutrients and can provide both a flavourful experience and a number of healthy microbial by-products. Even fermented soy germ can provide a wealth of antioxidants.

There is another soy derivative, but it's not for everyone. The clue is in the name: stinky tofu. This is a naturally fermented soybean product made with a variety of rotting and spoiling bacteria. Although there are some strains that can provide benefit to the body, many are also fecal in nature and essentially use up all the beneficial molecules before ingestion. Some people swear by stinky tofu, but for those with weak hearts, stomachs, or noses, it may be best to steer clear.

TAKE THE *SURSTRÖMMING* CHALLENGE Although meatballs may be the best-known dish out of Sweden, one of the locals' most treasured edibles is a fermented food called *surströmming*, or "sour herring." Anyone who has smelled or tasted this delicacy knows the name does not do justice. As a German food critic once put it, "The biggest challenge when eating *surströmming* is to vomit after the first bite, as opposed to before."

Fermentation is achieved by placing herring in wooden barrels full of brine and leaving them for three to four weeks. Bacteria break down the fish proteins and create a variety of by-products, some of which have an aroma. One of them, hydrogen sulphide, smells like rotten eggs. Another, methanethiol, gives off the aroma of rotten cabbage. (These are both toxins, but they pose no threat in very small quantities.) Once the fermentation process is over, the fish can be eaten immediately or put in cans for later. In the latter case, the smell can accumulate, as the bacteria are able to continue eating away at the flesh.

If you do choose to take the challenge, make sure your food prep area is well ventilated. When the can of *surströmming* is opened, expect to be overwhelmed by the odour. But it's worth the olfactory assault. The fish has a unique taste and will surely make for an experience you'll remember for some time.

THE FASCINATION WITH FERMENTED FISH The Swedish love for sour fish has gone global. All over the world, human cultures have turned to bacteria to deliver several unaesthetic, olfactory-offensive, and taste-bud-killing delicacies made from rotting fish.

The act of turning fresh fish into fermented delicacy requires expertise, a strong stomach, and most important, lactic-acid bacteria. They are the main drivers of the process, although they have to make do without their usual favourite foods, sugars and fibres. Instead, they feast on fish flesh as they break down the various tissues and turn them into easily digested proteins as well as a number of healthy fats, both of which help to regulate metabolism and digestion. Unfortunately, the bacteria also create an assortment of rather foul-smelling chemicals that may remind a person of a bad case of flatulence.

Fermented fish can be made more delectable by adding a small amount of sugar and fibre from a plant source, such as garlic. Both these elements provide an excellent source of nutrition for the species of lactic-acid bacteria known to make more satisfying flavours.

Fish sauce is the most common variety of fermented fish. Each country has its own version, but the process for making the sauce is essentially the same everywhere. Fish are cut up and then salted before being placed in a container with water and some form of sweetness, such as sugars, tree syrups, or cereal grains. The mixture is covered and then left to sit for a period of time ranging from days to months. The bacteria inside will use up as many of the nutrients as possible. But thanks to the sugars, the end products will be much more pleasant. The savoury sauce can be used as a seasoning or eaten by itself.

Fish sauce's biggest health benefit comes from the protein it contains. During fermentation, the fish tissues are broken down not only into individual proteins but into the most basic building blocks, amino acids. These are easily absorbed and can be readily used by the body for other purposes. Their presence also helps to lower the amount of acid and other enzymes needed to break down proteins. This can help to prevent acid buildup and reflux.

Then there are the by-products of the sugar. These include lactate, a major component of yogurt, and acetate, the principle chemical in vinegar. For the body, these chemicals help balance metabolism and work with the immune system to keep it calm. These chemicals can help to maintain the sense of fullness for a longer period of time. This can ultimately help in managing weight by reducing the urge for snacks between meals.

THE KOMBUCHA CONNECTION Take the dried leaves from a tea plant and add recently boiled water. Immediately, several beneficial chemicals are released, including antioxidants, anti-inflammatories, and even caffeine for those looking for a burst of energy. Add in some other plants, such as lavender or violet, and the drink becomes even more flavourful, as well as rich in healthy molecules to keep the body balanced.

But tea can be even healthier. Add some sugar and a number of microbes and eventually you get a tasty, fermented concoction commonly called kombucha. It's been a part of Asian culture for millennia. Usually made with black tea—although green tea will suffice—the drink is not only delicious but also imparts some very healthy by-products for both body and bacteria.

The process of kombucha fermentation is relatively simple. After the tea is brewed and cooled, some sugar is added to provide food for the microbes. Vinegar may also be added, as it has strong antimicrobial properties and will keep away unwanted microbes, such as pathogens. At this point, a combination of fungi and bacteria are laid on top of the mixture, which is then covered with a cloth. For the next few weeks (up to eight), the drink will ferment and the tastes will change. In the first few weeks, the flavours will be fruity and sparkling. If left longer, the mixture will sour, making the taste more challenging. But considering the benefits, a cup is well worth it.

Most of the sugar molecules are modified by microbes into a variety of by-products, which when ingested signal a state of calm to the gut. This helps to restore balance in the absorption of sugar so that the likelihood of spikes and crashes decreases. These molecules also signal the body that fat accumulation is not needed, preventing weight gain. This has several other indirect benefits, including better liver health, lower cholesterol, and normal insulin function, all of which help to prevent the onset of diabetes.

Antioxidants already found in tea are also modified so that their benefit is increased. In the digestion process, these molecules help to reduce the potential injury from acids and enzymes. They also protect the lining of the intestines so it's not prone to damage from high levels of oxygen, which can lead to cell death and ulcers.

One of the most intriguing benefits is the reduction in joint pain. During fermentation, a chemical called glucuronic acid is formed. When ingested, glucuronic acid is taken up by the body and made into an assortment of molecules known to help calm the joints. Two in particular, glucosamine and chondroitin sulphate, are so effective they are sold as supplements. But kombucha offers a natural source to allow the body to heal itself.

Though making kombucha is relatively easy, finding the required bacteria may be a challenge. For anyone wanting to try making the liquid treat, the best option is to find a starter kit with the right microbial population. This is called a symbiotic colony of bacteria and yeast (SCOBY), and it can be found in many natural health food stores. It will provide all the right species and ensure the tea will gain a pleasing collection of tastes and benefits.

THE ANCIENT TRADITION OF DRINKING BREAD The carbonated beverage kvass has been around for well over a millennium, and its name derives from an old Slavic word for yeast. It is made by fermenting bread, usually rye. As you might expect, the main microbial component is bread yeast, but there are also a number of lactic-acid bacteria. Each type of kvass has a different selection of microbes, and as a result, a different nose and taste.

The health benefits hailed by those who drink the concoction include improved digestion and elevated mood. These results cannot be explained entirely by the carbonation or the fact that most non-commercial versions of kvass have a small amount of alcohol (usually less than 1 percent). Instead, the answer lies at the molecular level. The drink has a high level of fibre, which can help friendly bacteria grow in the intestines. It also has high levels of minerals, vitamins, and amino acids—all of which are freed up from the rye during the fermentation process. These contribute to better health, and ultimately, a better state of mind.

If you want to try to make kvass, all you really need is some stale dark rye bread, water, sugar, and bread yeast to start the process. You could also add raisins and herbs like mint to give more depth to the flavour. Once you have all the ingredients, just follow these simple steps:

1. Toast the bread and then pour in boiling water. Let this simmer for a few hours as it cools to room temperature.

2. Add the raisins and mint at this stage if desired.

3. Once the liquid has cooled, add the sugar and yeast, and let the mixture sit overnight.

4. In the morning, strain the liquid through cheesecloth and store it in bottles capable of withstanding carbonation.

5. Give it a few days to settle and then enjoy.

I'll admit it's a unique taste, but I love it. I suggest the best time to drink it is in the afternoon on a warm day. You get all the taste and nutrients needed to keep you hydrated. Best of all, there's close to no alcohol so everyone can enjoy it.

DON'T OVERDO THE VINEGAR Apple cider vinegar is held in high esteem in the health-food world, and deserves to be. It is a hugely beneficial item. It's also incredibly simple to make. Just cut up or mash the fruit, put it into a barrel, and let it spoil.

Apples have their own bacteria and yeast populations, and these are quick to transform the delicious white slices into a mushy yellow slurry. First, the microbes use up the sugars from the apple flesh and turn them into alcohol (or more specifically, ethanol). This is apple cider, which for many makes for a great quaff.

To make the vinegar version, you'll need more time and another component: oxygen. Thankfully, oxygen is found in abundance in the air, so all you'll have to do is let the concoction breathe. The oxygen comes into contact with the ethanol and changes it to acetate, which is the primary component of vinegar.

Many vinegars are made from fruit using the same process. You can find vinegars made from pears, blueberries, mangoes, and of course grapes (in the form of wine vinegar). What sets cider vinegar apart from the others is the nature of the apple and its incredible benefits for us and our microbes.

Apples are nutritious for humans and for the bacteria in our guts. The fruit contains some simple sugars as well as a large amount of fibre. Although the human body cannot digest the latter, several species of bacteria simply love these molecules and use them for food. Their numbers rise, while those that cannot use these molecules—primarily unhealthy bacteria preferring high sugar and fat content—tend to drop in numbers.

Apples also contain a number of chemicals known as polyphenols. When the bacteria get a hold of these molecules, they transform them into antioxidants and anti-inflammatory molecules, which help the body stay balanced.

The most prevalent ingredient in apple cider vinegar is acetate, which is itself beneficial for health. It's a short-chain fatty acid and is one of the molecules needed to maintain proper metabolism. It can help to reduce hunger cravings and also regulate the levels of sugar in the blood to avoid spikes and crashes. The molecule can also keep a person feeling full for longer and stimulate fat burning. This means not only fewer feeding times but also increased weight loss. Although the effects are minimal in comparison to exercise and low-calorie diets, apple cider vinegar can help you shed those pounds. Acetate also appears to increase nutrient absorption. One of the most important increases occurs with calcium. Higher levels of calcium absorption can improve bone function and regulate blood pressure.

There is one more very important benefit to acetate: it's an antimicrobial. Vinegar has long been used as a disinfectant outside of the body, and it turns out it has the same effect on the inside. Acetate is able to either kill or stall the growth of several food-borne pathogenic bacteria, as well as yeasts and fungi. In the intestines, this can help control the rise of any uncooperative species wanting to cause trouble.

A drawback with apple cider vinegar is that it's easy to take too much. For most people, two tablespoons three times a day is plenty. Any more and there may be problems. Because acetate leaches calcium, the enamel on teeth may erode with continued exposure. Too much vinegar could also lead to a significant drop in blood sugar levels, which can in turn lead to low energy and even feelings of light-headedness. Also, potassium levels can also be affected as consumption rises. This can severely impact muscle strength and cause excessive sweating.

I love apple cider vinegar because of the way it helps me and my digestive system stay calm. I take a few tablespoons every night, with my smoothie. It makes for a relaxing drink before bed and keeps my gut balanced so I won't have to worry about any surprise intestinal movements in the middle of the night.

You can find apple cider vinegar in most natural health stores. It even comes in pill form for those travelling or simply unwilling to deal with the taste. The best products, though, are those that are sold raw (without pasteurization) and contain what is known as the mother. This cloudy slurry includes fibres left over from the fermentation process. Though we cannot digest these molecules, the good bacteria in our guts can and will happily enjoy the stringy treat.

THE END RESULT Here's an interesting task: head on over to an academic literature database like PubMed.com and type in the words *fermented, human, feces,* and *starter.* The number of articles published on the hunt for fermenting bacteria in excrement might surprise you (or revolt you!). Some of the titles alone are likely to cause a sudden lack of appetite.

It's a dirty job, but some microbiologists have to do it—and they have come up with some very interesting results. In many of the studies, bacterial species coming from feces performed exceptionally well as fermenters. They made the fermentation process go faster and produced a better overall product.

These discoveries also open the door to healthier fermented options. Advances in technology allow us to modify the conditions, the nutrients, and the species. This could lead to a new generation of fermented foods specially designed for a particular consumer. This personalized nutrition would represent a milestone in improving health, as it would ensure the most beneficial outcomes from a single food source.

It wouldn't be hard to achieve, either. The health-giving bacteria could be identified in the laboratory. Most would be lactic-acid bacteria, which are the primary types used for fermenting and also tend to provide the body with the most benefits. Once isolated and growing, the bacteria could be used to ferment a nutritious source of food, whether dairy, vegetal, or grain. The result would be a personal fermented food designed to be healthy to the individual.

One added advantage is that the strains could be frozen for later use. The bacteria in the gut can suffer severe consequences should an infection occur requiring an antibiotic prescription. The diversity of the microbial population could suffer and, as we've seen, potentially lead to the onset of chronic health problems. Should this happen, the frozen strains could be woken up from their slumber and given back to the person to restore the microbial balance.

THE TROUBLE WITH PROTEIN All life needs proteins to thrive. They are needed for proper cellular structure and function, and even to help us recognize various signals coming from the outside. Chemically, a protein is a chain of molecules known as amino acids; they are the basic building blocks of life. But these molecules are not readily available as individual entities. To get them, an organism needs to ingest proteins from another source and then find some way to break down the larger molecule into more useful pieces. Both humans and bacteria have enzymes to target protein bonds and shorten the chain. Eventually, the individual amino acids are freed and become available for the construction of different proteins.

Almost any natural food source contains proteins. The most abundant source of amino acids is meat, which is mostly made up of muscle, the most high-protein part of the body.

Only twenty of the five hundred or so known amino acids make proteins. The rest are employed in cell-to-cell signalling and the production of energy and waste. These other compounds may not be as useful to the body, but can be difficult to avoid when eating a highly complex food such as meat. Unfortunately, when taken into the body, these compounds can lead to a higher risk for chronic health problems.

One such chemical is L-carnitine. This molecule is responsible for the production of energy from fats and carbohydrates. It's made in the liver and kidneys of animals, and is important to humans in maintaining proper metabolism. But bacteria don't find this molecule particularly useful and tend to break it down into a waste product similar to those found in rotting fish. Most of the waste will eventually end up in the feces and be eliminated from the body. But some will be reabsorbed into the gut lining and transferred to the liver. The liver can use these waste products to determine whether the gut needs detoxifying. If it does, the liver then changes the structure of the toxin so it can be eliminated in the urine.

This is when the trouble starts. While making the journey from the liver to the kidneys, this now-modified toxin can interact with other cells and inadvertently signal that the liver is stressed and needs help. The body will respond to this signal by reducing the amount of bile transported from the liver to the gut for digestion. This causes an overabundance of bile in the liver and stops the manufacture of the basic building block, cholesterol. Over time, cholesterol will find its way to the arteries, where it can cause atherosclerosis and other cardiovascular diseases.

The risk of atherosclerosis from red meat hinges on the bacteria in the gastrointestinal tract. A diverse population reduces the levels of toxin production (only a small number of bacteria are capable of creating a toxin from L-carnitine). The risk rises when the diversity is tipped in favour of toxin producers.

Bacteria that prefer sugars are the most likely to cause problems. They take in protein slowly and have no need for the massive rushes offered by red meat. When they come into contact with L-carnitine, they convert it into the toxic waste product and get rid of it without any fuss or bother. The toxin ends up in the intestines, where it can be absorbed.

L-carnitine is found in all meats, but red meats have the highest level. This doesn't mean red meats should be avoided altogether, but they should be eaten in moderation, as part of a sensible diet. Just remember to think where you might be a few hours later. The heavy and dense nature of the meat requires the intestines to stall digestion. This is accomplished by producing very small amounts of hydrogen sulphide gas. For the intestines, this chemical signal ensures that all the nutrients are properly digested. But once the hydrogen sulphide is released from the body as flatulence, the entire area is engulfed in the smell of rotten eggs.

CHICKEN SOUP FOR THE COLD When I've got a cold, I avoid over-the-counter medication and instead head to the meat section of the grocery store. For me, a chicken provides more benefits than any pharmaceutical. It's a natural way to deal with the sniffles, coughs, and sneezes, and it improves my health sooner rather than later.

Chicken soup contains vitamins to keep the metabolism strong, antioxidants to prevent the stomach from getting upset, and anti-inflammatories. But the entire mix has to be boiled so these can become readily available in the broth.

The effect of eating chicken soup is twofold. First, the raging immune system is tempered. This can reduce fever, aches, and pains. It also helps to focus the defences on the job at hand, which is to take care of the viruses infecting the body. These pathogens have a habit of sending the immune cells on wild goose chases, leading to far more damage than necessary. The soup can counteract this by signalling the body to forgo any unnecessary actions.

Second, it helps control the symptoms associated with the condition known as the "man cold." This combination of anxiety, depression, and a sudden lack of independence is caused by stress molecules, and it forces the poor male to seek attention and be nursed. It's purely psychological and has no chemical basis; there is no difference between the way men and women respond to a cold. Yet chicken soup can help reduce the levels of these stressors and at least bring some comfort to the mind.

If you need chicken soup for a cold—or indeed the soul— the best option is to make it yourself. This requires more work than grabbing a can or package off the shelf, but it can be far more beneficial. Most processed soups have higher levels of sodium and other ingredients that you really don't need. Making soup at home allows you to control these ingredients and determine how much you'll use. You'll also benefit from the aroma as the chicken soup boils. Even the steam can help to relax the mind and bring a sense of calm while you convalesce.

DIGEST AT LEISURE You've no doubt heard the stories about turkey and sleepiness. This apparently has to do with the amount of a chemical known as tryptophan. It's an amino acid and a building block for our proteins. It also serves as a neurochemical, making the brain feel tired and in need of a nap.

It's a great theory, but it's not entirely true. After all, there's a higher concentration of tryptophan in chicken and even soy, the base for the meat-substitute tofu, and they don't have the same effect. So what's causing that need for a nap? The answer may have more to do with microbes.

When you eat a turkey dinner—or any large meal, for that matter—you are taking in a great number of calories. The bacteria are more than happy for the surge in nutrients. They work away at both eating and multiplying their numbers, not to mention the number of waste products. Some are toxic, while others are inert. But some, such as the molecule butyrate, or the amino acid tryptophan, are recognized by our cells as calming and can increase the levels of a number of good brain chemicals. When the brain senses what's happening in the gut, it realizes all is good and the body should rest. That translates into drowsiness.

This microbial messaging is a perfectly natural phenomenon, and it's good for us in many ways. When we are digesting, the last thing our bodies need is exercise. Rest allows for better digestion and keeps the focus of the metabolism where it belongs: in the gut.

Once digestion is over, several hours later, we can return to our normal activities and use up some of that excess energy. In other words, going for a walk right after dinner may not be a great idea, but a jog the next morning is perfect. The bacteria will stay content for about half a day, but soon after they'll start to look for more food, preferably of the same size and density. To keep them in line, the next meal should be lean and filled with fibre.

PHYT FOR YOUR LIFE Plants make up a significant portion of any healthy diet. They are chock full of sugars, proteins, vitamins, minerals, and even some fats. But contained within vegetables, fruit, and beverages such as coffee, tea, and wine are a group of chemicals known to provide both body and bacterium with health benefits. Collectively, they are known as phytochemicals.

At the microscopic level, these molecules are highly diverse. Many are colourful, while others can trigger a taste response. Flavonoids, for example, give off hues ranging from blue to yellow to purple to red. Tannins are easily recognized by a sharp flavour on the tongue; they add depth to several foods. Alkaloids like caffeine are to some people unacceptably bitter. Then there are compounds that give off heat, such as capsaicin in chili peppers, and the nasal-opening sulphur chemicals found in garlic, onions, mustard, and wasabi.

Although bacteria cannot taste or smell, they have the ability to recognize these molecules and react accordingly. Many species, including some known to originate with plants and soil, thrive on these chemicals. These microbes get into us by way of the raw vegetables and fruit in our diet. In the presence of their preferred nutrients, they can make a home inside our guts. There they contribute to increasing the diversity and help to fight off any of those pathogenic foes. This happy state continues for as long as our diet includes these plant-based foods.

The benefits of phytochemicals continues to grow but two are hailed as effective to improve health. The first is the state of physiological calm. Phytochemicals from green tea, fruit such as grapes (including their fermented counterpart, wine), and spices such as turmeric provide the body with antioxidants and can have the effect of calming anxiety. But that is not the only source. These chemicals also prompt the good bacteria to make other calming agents. The body takes in these chemicals and spreads relaxation to the immune system. Inflammation is reduced and a more natural state of metabolism is restored. Over time, the diversity in the gut shifts as those bacteria unable to deal with these chemicals either halt their growth or die off, leading to less likelihood of an inflamed state.

This leads to the second benefit of phytochemicals, weight loss. With inflammation down and the pathogens leading to weight gain reduced or gone, the body can naturally start to trim down. The bacteria and their metabolites help by restoring the natural balance of hunger and food intake. Without the potential pathogens, the need for sugar lessens, as do the chances for sugar highs and lows. This leads to reduced sugar and fat cravings and the re-establishment of a more routine eating schedule.

Now weight loss can truly kick in as the formation of fat tissue due to inflammation and poor diet is slowed. The process may take longer than fad diets, but it's more likely to last. The goal is to introduce good chemicals through food rather than to starve the body with dietary restrictions.

THE GIFT OF GARLIC Garlic is one of the healthiest foods we can eat. Each clove is rich in fibre and offers a wealth of nutrients to friendly bacteria. But this isn't the major reason for chewing on the fibrous mix. On the inside is a chemical called allicin, an antimicrobial molecule known to help garlic defend against pathogens in the soil. When eaten by us, it has a similar effect, killing off bacteria, particularly unfriendly ones. It's also effective against viruses such as the flu.

But that's not all that's good about garlic. In the gut, allicin has anti-inflammatory properties. This is particularly helpful if any of the unfriendly bacteria are attempting to overgrow and attack the intestines. Allicin can break up their effort and send the bacteria out of the area, bringing an end to their harmful campaign. The chemical can then help to soothe the intestine by promoting the formation of antioxidants and anti-inflammatory molecules. Although it may not eliminate the bad bugs entirely, it can definitely help to keep things balanced.

If you're eating garlic, there is a trick to keeping your breath from turning foul. Within the clove is a stem; this is where the smell resides. Take it out and you will be able to enjoy the benefits without having to worry about the lingering olfactory effects.

THE ANATOMY OF FLATULENCE The gas in our bodies has to come out somewhere, and in the intestines there's really only one exit. We may be able to get away with the odd odourless release, but more often than not, we have to deal with a pungent smell and the social consequences.

Gas formation is a natural part of digestion. Air is taken in when we eat, and if it's not immediately regurgitated in the form of a burp, it flows through the gastrointestinal tract until it reaches the other end. As we digest our food, all the nitrogen, hydrogen, oxygen, and carbon dioxide molecules produced by the body are also collected here for later expulsion. This makes up about 40 percent of the gas we form.

Most of the other 60 percent comes from those trillions of bacteria living in the gut. They too digest food and make several different by-products from the nutrients. In particular, they make three molecules. One, methane, is odourless. But the other two—methanethiol and dimethyl sulphide—contain a certain element known to tickle the nose: sulphur. They produce aromas that, not surprisingly, smell like feces.

There is one final flatulence fragrance, and it is by far the worse: hydrogen sulphide, or "rotten eggs." Our noses can pick up incredibly small concentrations of this chemical, so any emission is guaranteed to be detected. Bacteria make it as a waste product, but they are not the only producers. We use it as a way to slow digestion of dense foods like meat and eggs.

Avoiding gas really isn't an option, but we can reduce the chance for a smelly emanation. All we have to do is avoid dense and sulphur-rich foods such as those meats, eggs, beans, cabbage, and Brussels sprouts. If this isn't an option, you can purchase supplements to help minimize the amount of flatulence produced. These pills contain enzymes made naturally from microbial species and are perfectly safe to take. They can help you enjoy the fun of a social gathering without having to worry about embarrassment later on.

CLEAN AIR ACTION There is another answer to anti-social smelly flatus. It's called activated carbon. It looks like a solid black grain, but when viewed at the microscopic level, it has an incredible amount of surface area. It's formed through the charring of plants such as wood so that the cellular material is destroyed. After the charring process, the black substance, carbon, is able to bind and trap molecules of all shapes and sizes. This makes it perfect for filtration, as it will remove a variety of chemicals from air, water, and gas. In the laboratory, activated charcoal is the perfect way to prevent the air from becoming foul. There are two ways to use it. The first is to ingest it, usually in pill form. It is safe in small quantities and can help remove not only odours but also potentially toxic chemicals such as metals and alcohol; a pill may even be able to stop a hangover. But this doesn't always guarantee an odourless emanation.

The other way is to put activated carbon near the rear end to catch those smells while they're being expelled. The best way to accomplish this is with underwear that has the carbon meshed into the fabric. This can definitely help anyone with concerns for social embarrassment. You have to make sure the underwear is snug, though, so that the gas goes directly into the carbon.

Entamoeba histolytica

8. DIET

SO SAD Weight gain is commonly thought to be caused by overeating. But more often than not, the problem lies not with the quantity of food but with its quality. In industrialized countries, where fast food is rampant, obesity rates are, not surprisingly, the highest.

In the United States, the average diet is not just poor, it's atrocious. Health officials have even given it a name, the Standard American Diet, or SAD. SAD is primarily made up of foods containing high amounts of simple sugars, such as glucose and fructose, and saturated fats. Both of these are necessary for proper health, but their levels in the diets of most Westerners (SAD is also known as "the Western Diet") are so high they interfere with the body's usual metabolic processes. If that body is also insufficiently exercised, the deposits become larger and so does the body. This isn't the worst of it. In the intestines, the bacteria that prefer this diet tend to be foes. The increased levels of sugars and fat let them thrive.

These bacteria are the same ones that can influence disease. As their numbers grow, the outlook for us becomes increasingly dire.

The most troublesome microbial consequence of SAD is lipopolysaccharide (*lipo-* meaning "fat" and *-polysaccharide* meaning "starch"). LPS, as it's more commonly known, is a single molecule found on the surface of a class of microbes called Gram-negative bacteria. The name comes from a test performed in the laboratory called a Gram stain, named after its inventor, Hans Christian Gram. Under a microscope, the microbes appear either purple, known as Gram-positive, or red, Gram-negative. Gram-negative bacteria include several infamous species such as *E. coli, Salmonella, Neisseria* (as in gonorrhea) and *Pseudomonas aeroginosa* (the wound infecter).

LPS is a mixture of two different chemicals, sugars and fats, and helps to keep the outside wall of the cell intact. Much like our skin cells, when this molecule has served its purpose it flakes off and disperses. But, unlike our skin cells, which simply become one with the dust, LPS tends to stick around in the intestines and make contact with other cells, including those in the intestines. This is where the trouble begins.

The cells of the intestine have sensors such as the toll-like receptors I discussed in Chapter 1. They detect LPS, and offer an early warning system for an impending invasion. Many bacteria species that cause diarrhea release LPS as they get ready for an attack. As soon as TLRs spot the toxin, they call for the immune system to mount a defence. Even if no attack comes, the gut cells cannot discriminate between a bacterium enjoying a meal and one that intends to invade. It's an evolutionary error on the part of our immune systems, and it can have rather grim consequences.

For one, the cells of the intestine become more porous, allowing passage of LPS into the bloodstream. Over the course of a few hours, the molecule can make its way into the blood, provoking the entire body's immune response as opposed to a particular region. But this isn't going to cause traditional symptoms of fever, aches, pain, and fatigue. Instead, when this molecule moves through you, it may cause something completely different: hunger.

Here's what's going on. Several hormones control our appetite, including one that comes directly from the stomach. It's called ghrelin (*ghre-* meaning "to grow"), and the stomach secretes it in order to notify the brain it's ready for some food. When LPS gets into the blood, ghrelin levels shoot up as the body looks for more energy to sustain a defensive posture. The cravings are short-lived, and the levels decrease a few hours later. Yet they can be so strong at their peak that a person just has to binge-eat.

For people on a diet, whose goal is to resist these cravings, this can be a problem. Eventually, the ghrelin levels will fall. But because the cravings were not satisfied, another hormone is triggered: cortisol. This is the stress hormone, and it forces the body to preserve energy in the form of fat, particularly in the abdominal area. Over time, this form of ghrelin-starvation leads to weight gain.

The best way to avoid this cycle of inflammation and weight gain is to stay away from the Standard American Diet altogether. If that is too radical, SAD can be cheered up immediately by the addition of fibre, which keeps the gut wall free from LPS, and so helps the intestinal cells to maintain their tight fit. This leaves LPS with fewer opportunities to access the blood. It's not foolproof, mind you, but regular intake of about 40 grams of fibre a day can not only lower level of LPS in the blood but reduce hunger pangs and keep you feeling sated for longer periods of time.

THE TROUBLE WITH GLUTEN Wheat has been a major part of the human diet for close to ten thousand years. Along with barley and rye, it has become a staple in many parts of the world and makes up close to half of all calories consumed. But millions of people around the world eschew these grains for fear of intestinal pain, nausea, diarrhea, skin rashes, and even breathing problems.

The source of these wheat-related issues is a compound known as gluten. This sticky substance gives structure and texture to bakery products as it helps dough to rise. At the molecular level, gluten is made up of a number of proteins called gliadins and glutenins. These are extremely resistant to heat, making them perfect for baking and other high-temperature food manufacturing. They are also hard to digest, as they have several chemical bonds resistant to breakdown. In humans, only one enzyme in the gut is capable of handling these proteins. Unfortunately, it's not entirely effective, and many gliadins remain untouched. As these proteins accumulate, the immune system sometimes gets involved, sensing a foreign presence and classifying it as a toxin.

When undigested gliadins are detected by the immune system, there may be a rather nasty consequence. Pain, nausea, and diarrhea are the first symptoms as the brain and nervous system are triggered to get rid of the unwanted intruder. If the proteins are not entirely eliminated, the immune response goes systemic as the body searches elsewhere for trouble.

The most obvious location is the skin, and so the immune cells and antibodies head there to look for the enemy. As the skin becomes flushed with these cells, rashes may develop. In the respiratory tract, a rush of immune cells may lead to irritation and breathing troubles.

If the immune response remains active over a long period of time, other parts of the body may also be affected, including the brain. Higher levels of anti-gliadin immune agents have been noted in people suffering from autism spectrum disorders. This doesn't mean gluten is a cause of autism, but several clinical trials examining gluten sensitivities and autism spectrum disorders suggest the symptoms of autism might be worsened by gluten and improved with its removal from the diet.

Conventional wisdom states that to avoid these many potential hazards hazards of gluten, it has to be avoided completely, with all the vigilance and lifestyle changes that involves. But for the 93 percent or so of the population who are not genetically or immunologically unable to eat gluten, there is another answer. It lies with the microbial population in our guts and mouths. Many species living there are able to digest grains and their proteins, including gliadins. They do this by producing enzymes that specifically target those tough chemical bonds and break them with ease. These bacterial enzymes are extremely efficient and can completely eliminate the gliadin threat.

The problem is that those who suffer from the various complications associated with gluten tend to have fewer of these bacteria in their bodies. This lack of proper diversity, sometimes called dysbiosis, allows the gliadins to remain untouched and forces the immune system to act. Restoring these bacteria to the gut is not easy—it may take years to re-establish the enzymes needed to properly digest these proteins.

There is an easier way: fermented breads like sourdoughs and Asian flatbreads such as dosa. Grain fermentation lets the bacteria destroy the gliadins before they are ever ingested. Those same lactic-acid bacteria we consider to be friends can be used to break down the gliadins in the dough. These bacteria also stimulate the grains themselves, encouraging them to become active and break down the proteins. The end product is a little less stretchy and has a sour taste—due to the production of lactic acid and other metabolites—but the gliadins are gone, making it safe for almost everyone.

The art of souring dough is still practised in most bakeries. It's not hard to do at home, either. All that's needed is flour and a starter culture, which can be bought from your local sourdough baker. Or if that's not feasible, use plain yogurt from the grocery store. It contains the same bacteria. As for a recipe, there are plenty on the web just waiting to be tried.

INFORMATION LEAK Leaky gut syndrome occurs when the normally solid and impenetrable intestinal wall develops microscopic holes, creating a two-way street between the intestines and the rest of the body. You may think this is a bad thing—after all, who wants feces in the blood or vice versa?— but it is actually a natural process. It is a defence mechanism to increase the amount of water in the intestines. The goal is to flush out any molecules the body senses as toxins, such as undigested sugars, starches, alcohol, and fats. It's not a fun time—this syndrome may be accompanied by pain and cramping, as well as a rush of diarrhea. But usually, everything soon returns to normal.

For some people, however, the floodgates are kept open for much longer, and this allows toxins and bacteria to make their way upstream and into the intestinal tissue and the bloodstream. Most of the time, the body's defences are able to handle any unapproved entries. But if the holes persist, the amount of bacteria and other unwanted molecules can rise and overwhelm the troops. This can make life unpleasant, with symptoms persisting for months, if not years. And larger problems can also develop, including insomnia, malnutrition, memory loss, and even the onset of severe diseases such as multiple sclerosis, arthritis, and colon cancer.

One cause of long-term leakage is the presence of unfriendly bacteria wanting to head into the body. But they are not the only culprits. Some food ingredients have the same dam-busting power. The most well-known of these is undigested gluten.

Once gluten is ingested, it travels to the gut and clumps together. The intestines will try to digest the various proteins, but they may have trouble breaking down the mass. The solution is to add more water to the mix, and so the gut is told to open up and flood the area. When this happens, the gluten will be forced further down the gut, ending up as either bacterial food in the colon or a rather unpleasant fecal output.

For people suffering from celiac disease, intestinal leakage due to gluten is common and essentially means they must avoid the substance altogether. The same seems to be the case for some sufferers of other intestinal disorders; removing gluten can help prevent pain and other symptoms.

For you and me, intestinal leakage is usually just a part of everyday life. That being said, keeping undigested gluten to a minimum is probably a good thing. There's no reason, though, to go completely gluten-free. Unless you have an allergy, intolerance, or celiac condition, a little gluten in your life is nothing to worry about.

CARBS HAVE THEIR USES People who are trying to lose weight usually subscribe to a well-known mantra: cut down on the carbs. Reducing the level of carbohydrate intake does have its value, as it decreases the flow of easy-to-digest sugars and forces the body to find alternative sources in adipose tissue, also known as our fatty zones. But some people go a little too far by rejecting carbohydrates completely and switching to meats and fats. This may cause more harm than good.

When sugar is in short supply, organisms need to find alternative energy sources. Some species are unable to thrive without it and will not stick around in a gut devoid of sugar. The rest will feed on whatever they can find, including fats and proteins. The resulting changes in the microbial population can have a dramatic effect on the body. When bacteria break down proteins, they create a number of potentially unhealthy waste products, including ammonia, a known toxin, and several molecules known as nitrosamines. The latter are smelly compounds associated with rotting meat, and they can damage human cells. Most of these molecules are sent out of the body in the feces, but some end up in the blood. Once there, they cause significant injury and even death to human cells. The most frequently damaged organs are the liver and the kidneys but the gut itself is also a target. These molecules may also increase the risk of colon cancer.

But the effects don't end there. One of the hallmarks of nitrosamine exposure is the development of chronic diseases. The toxin causes the body to go from a state of balance to a state of stress and slow down the metabolism to preserve energy for a fight. This sparks an imbalance in the levels of sugar in the blood and triggers insulin resistance, the first sign of Type 2 diabetes.

The problems can also migrate to the brain, where a similar reduction in cellular activity occurs. When this happens, the cells become more vulnerable to starvation and eventually death. In a short-term situation, the effects are transient and can be repaired quickly. But consistent exposure to low levels of toxins can starve the cells of energy and cause them to die off.

You can reduce nitrosamine transfer to the body by increasing the number of chemicals known to keep its absorption to a minimum. But these chemicals are not produced by the cells of the intestines; instead, they are formed by several species of bacteria. The only way to get them is to increase your intake of foods they like, which include carbohydrates and fibres.

Essentially, the way to prevent the problems associated with a high-protein, low-carbohydrate diet is to give up the diet itself.

That doesn't mean all carbohydrate-restricting diets are doomed to fail. But if you want real, lasting weight loss, you should forget about removing all carbs and reduce the fat instead.

Normally, humans ingest about 15 percent of their calories from proteins and about 35 to 40 percent from fat. If you change those values to 30 percent protein and 20 percent fat, without taking away any of the carbohydrates, weight loss will happen.

There are two reasons for this. First, fat has many more calories than sugar, so when fat intake is reduced, a net loss in calories occurs. Second, lasting weight loss relies on bacterial diversity in the gut, which will remain more balanced with the aid of carbohydrates, particularly fibre.

BACTERIA LOVE VEGETABLES If you're trying to cut meats or dairy from your diet, you know it can be a struggle, both socially and physiologically. And those who go to even greater extremes—following a vegan diet, for example—can miss out on the nutrients needed for optimal health. These include vitamins D and B12, as well as calcium and zinc. These four are important to proper metabolism, immune function, and psychological balance. But they can easily be provided through supplements or nutrient-fortified food products.

Bacteria have no such concerns. They simply love vegetarian diets. Fruit, vegetables, and legumes provide all the nutrients bacteria need for optimal growth. As they eat and digest, their numbers grow exponentially and their diversity increases too.

Each one can then contribute metabolites, not only to improve digestion but also to maintain the health of the colon and other parts of the body.

A vegan diet also creates greater acidity in the colon. As the bacteria chew on fibre, they release a number of different acids. For an omnivore, this acidity is balanced out by the buffering capacity of milk and meat. But when these buffers are absent, the pH level goes down. Although acid in the gut is considered a bad thing (think of acid reflux), it is quite beneficial in the colon.

Many bacteria known to cause illness cannot survive in a low-pH environment. Their numbers will drop, and in some cases, they'll disappear altogether. This means a lower risk for infection and chronic ailments such as diarrhea and intestinal pain. For example, a shift to a vegetarian diet can reduce the levels of bacteria known to exacerbate arthritis or contribute to colon cancer. Although this isn't a perfect means of prevention, it's worthwhile for those at risk.

Vegetarian diets have another bacterial benefit: the production of a metabolite similar to estrogen. It's called equol, although this has nothing to do with equality; it was first found in horse urine. When equol circulates in the blood, it has two major functions: it acts as an antioxidant to prevent any cellular damage during times of stress, and it regulates the levels of sex hormones such as testosterone and estradiol (which can lead to breast cancer when in high concentrations). Equol helps prevent estradiol excess by essentially first blocking its activity, escorting it to the liver for processing as waste and eventual disposal.

When other foods are restricted, equol has a better chance to get into the body. This in itself may make a vegetarian or vegan diet seem like a good idea, microbially speaking.

The change to a plant-only diet does have potential drawbacks. Some friendly bacteria do end up losing ground as they go without their most valuable resources, meat and milk. Some of these bacteria have been with us from infancy, having been introduced in breast milk and maintained over the years. Others are more common to the colon and have established a balance with the body. Their departure makes the journey towards health more difficult as a result of improper energy sharing.

Many of these bacterial strains are known for sharing energy reserves with their human counterparts. This means less efficient digestion, which in turn can lead to weight loss and lower metabolic function. Without proper management of weight loss, vegetarians or vegans can easily become unhealthily underweight. This is a particularly important consideration for athletes, who expend far more energy and need to recoup as much as possible. More frequent intake is needed, with a focus on simple sugars from fruit. Adding some animal-based products into the mix—so-called ovo-lacto vegetarianism— can only make for an even healthier life.

GO WITH GOAT Milk might just be the world's best food. It contains all the sugars, proteins, fats, vitamins, and minerals necessary to keep the body nourished and healthy. But the benefits vary according to the source, which can be human, cow, goat, and even sheep.

For humans, as I discuss later, there is no better milk than human breast milk. The nutritional value is only part of the advantage. The rest comes from the inclusion of a number of factors from the immune system. These include cells to identify friends and foes, antibodies to prevent damage from foreign entities, and antimicrobials to keep pathogens from causing undue harm.

A molecule called lysozyme specifically attacks bacteria known to cause diarrhea and other gastrointestinal problems. The concentration of this molecule in human milk is up to three thousand times higher than in milk from cows, goats, or sheep, and as such can act as the perfect means for microbial population control. Another molecule, known as lactoferrin, works with the immune system to target, attack, and kill pathogens. In human milk, this molecule is at levels about one hundred times higher than those produced by animal milk.

Human milk is best when sourced from an individual's mother. Each milk is specifically geared to nourish her own offspring. So for those of us who are not infants, thank goodness for cows and other milk-yielding animals.

While drinking milk does a body good, it's the Wild Wild West for the bacterial population. In the gut, all bacteria can use the nutrients, so those that have the competitive edge— whether through faster capture and harvesting of those nutrients, or through microbial warfare—will get the lion's share. This usually means that without some type of control mechanism, like lysozyme, the good bacteria will inevitably lose out to potential pathogens. They are simply no match for species tailored to grab nutrients, grow rapidly to overpopulate the community, and then release toxins and other agents to kill off competitors.

When cow's milk is consumed regularly, a person's bacterial population may become imbalanced over time, and this can lead to shifts in the body's ability to interact with the bacteria and even the milk products themselves. Milk-associated problems include lactose intolerance, pain in the intestines, and significant weight gain.

Perhaps the cruellest consequence is the onset of an allergy to milk itself. This comes about when the immune system cannot deal with certain proteins without the assistance of bacteria capable of breaking down these proteins into harmless by-products. In this state of imbalance, where these species are absent, an allergy can quickly arise, usually in children, but sometimes in adults.

All things considered, it is perhaps time that the cow gave way to the lowly goat. It's true there's already an entire branch of the dairy industry devoted to the goat, but it pales in comparison to the near-monopoly of cow-based options. Yet a change may one day be in order.

The fat content in goat's milk more closely resembles that of its human counterpart, but goat's milk doesn't have as much sugar, which is good news for anyone counting calories. The calcium, magnesium, and potassium levels far surpass those of the cow, making goat's milk even healthier in terms of trace nutrients. Then there are vitamins A, E, and K, which are present at higher levels in goat's milk and align with breast milk.

There's another advantage to goat's milk—it tends to act like human milk in keeping the bacterial population balanced.

The lower sugar content and higher fat content contribute to that balance, but the key factors are the lysozyme and lactoferrin. Although the levels of both are much less concentrated than in human milk, there appears to be enough available to help control the bacterial population in the gut. The overall result is a microbial population resembling that of a person who has been fed human milk.

BEYOND THE PALEO You'd think that with all the years that have passed since the days of the cave dwellers, our bodies might have adapted to a different diet. But many people swear by the Paleolithic (or Paleo) diet, which is based on eating the same way our hunter-gatherer ancestors did thousands of years ago. This means lots of fruit, vegetables, nuts, seeds, plant oils, grass-fed meats, and seafood. There is no room for processed foods or any agricultural grains such as wheat or legumes (although oatmeal was part of some ancient European diets).

From an evolutionary perspective, the Paleo diet has merit. The genes that make us human have changed little over the last ten thousand years. Because our bodies are relatively unchanged, this diet should be good for our health. But there is another factor to consider: for the bacteria living in our gut, times today are very different than they were back then.

The microbial population, especially inside people living in developed countries, has undergone a profound change, thanks to a transformation in the human way of life. Humans are now less nomadic and tend to stay in one area for long periods of time. They also choose to eat foods produced by way of agriculture and animal husbandry. The effect of these changes on our microbial populations is probably best illustrated by looking at present-day populations living outside the modern concept of civilization. Take the Hadza people, for example. They live in a small area of Tanzania, and many still survive as hunter-gatherers. The men go out looking for food while the women stay back and tend to the daily chores, including gathering water. The bacterial population in these individuals is quite different from that of someone living in a city. These people even carry microbes never seen in modern human civilizations. There are also differences between the men and the women in the same village. The women have bacteria more aligned to eating plants and tubers; the men have bacteria more closely related to meat and the primary source of sugar, honey.

Another Paleo population can be found in the Amazon jungle. They are the Yanomami Amerindians, and they have remained largely untouched by civilization for close to fifteen thousand years. Unlike the Hadza, they have developed a modest form of agriculture, although they are still for the most part foragers. They too have a very different population of bacteria compared to people in developed nations. Their bacterial population is also incredibly diverse, with many more species found within the intestines and on the skin. Also, because they all eat the same food, everyone in a particular village has a similar bacterial profile.

This diet and the resulting bacterial diversity bring many benefits. The metabolism of nutrients is incredibly high, increasing the amounts of amino acids, vitamins, and minerals. The formation of energy is also much better, meaning the body is able to live off less. Then there are the phytochemicals, which can bring the body into a general state of calm. The Paleo lifestyle is, as claimed, good for health and good for overall well-being.

But there is a catch. This incredible bacterial diversity among the true Paleo people can't be achieved just by way of diet. Another important source of the bacteria is the environment. To get the true benefits, you would have to live in a jungle or on the plains. Otherwise, the Paleo diet may not do anything for your bugs or you.

THE FODMAP SOLUTION Diets come in all shapes and sizes, and many have inventive names. You've probably heard of the South Beach Diet, which essentially eliminates all carbohydrates. Maybe you've come across the chicken soup diet, which, as its name suggests, has you eating nothing but chicken soup for a week. My personal favourite is the chocolate diet, although I'm not sure it is altogether effective. One thing is for certain: if a diet has a catchy name, there's a good likelihood it's a fad and won't last.

There is, however, an exception. It's called the low FODMAP diet, and it was first proposed in 2005 to help people suffering from Crohn's disease. The idea is to reduce the intake of so-called FODMAP foods—fermentable oligosaccharides, disaccharides, monosaccharides, and polyols. These foods may be fermented by bacteria, but they are poorly absorbed by the body, leading to gut permeability (aka leaky gut syndrome). The FODMAPs include foods such as corn syrup, ice cream, wheat, onions, lentils, soybeans, and artificial sweeteners. (The entire list of FODMAPs can be quickly found online.)

The low FODMAP diet is a step above the rest because its benefits are known and well tested. Removing foods like onions and lentils will improve overall absorption in the gut. This means less food for bacteria, particularly those known to cause gas and toxins. This not only saves the intestines from pain but also keeps bowel movements firm and less smelly.

The low FODMAP diet has become an effective treatment for people who suffer from problems in the gut, including irritable bowel syndrome and other gastrointestinal disorders. But even if you don't have those issues, you can benefit from keeping those FODMAP foods to a minimum. This could also be a nice way to avoid some foods you don't like to eat, such as cabbage, broccoli, avocado, and asparagus. But don't think you can get away with saying no to eggplant or carrots. They are excellent sources of nutrients, and they have very little FODMAP content.

THE MICROBE-FRIENDLY DIET Most diets deal with *loss* (usually weight), but there is one that focuses on *gain* (in friendly microbes). It's the Mediterranean diet. The diet has a simple premise: every day, you should eat a combination of legumes, seeds, fruits, vegetables, and non-refined grains such as rice. A few times per week, you are allowed dairy products, fish, polyunsaturated oils such as olive oil, nuts, and fermented foods. Red meat is restricted to a few times per month. Wine can be had daily, but in moderation.

The vegetables, fruits, and nuts provide fibre and numerous phytochemicals to bulk up the digestive tract and keep it running smoothly. The oils and fish provide the fats needed to keep the cholesterol balanced, and the grains are an excellent source of carbohydrates. The milk and dairy provide calcium, vitamins, and minerals, while the meat rounds out the benefits by forcing the liver to make bile, and so giving it a workout.

The microbial populations of people who follow the Mediterranean diet are also likely to be in perfect harmony. There is something good in this diet for all bacterial inhabitants of the intestines. Those that love simple sugars are given their fair share. The fibre lovers are kept happy with a plentiful supply. Meat levels are balanced so that proteins can be broken down without producing high levels of toxic by-products. There's even something for those species looking for fat.

A balanced microbiological population can also help to stave off a number of diseases. Conditions such as obesity, diabetes, cardiovascular disease, allergies, arthritis, liver and kidney failure, and even some psychological disorders have a link back to the gut microbes. Usually, the problem is imbalance—in which one population overwhelms others, leading to a condition known as dysbiosis.

But perhaps the greatest benefit comes from what the diet doesn't do. Most diets have a tendency to shift the bacterial population towards one type of species. The only shift seen in the Mediterranean diet is towards the prevention of pathogenic organisms that may cause harm. Many of the species already in the body are able to produce antimicrobial chemicals to keep invading species at bay. The likelihood of gastrointestinal distress is therefore minimized.

Although this diet may be the best for microbial balance, it is not intended to cure anything. It won't help to resolve arthritis, nor will it cure cancer. But it can help to reduce the risk of the onset of disease and may have preventative value for those who are on the verge of a less-than-healthy way of life.

This diet may help vulnerable human populations, such as the extremely elderly, stave off chronic diseases. It may also help those who are at risk of variations in their microbial populations because of travel. Maintaining the bacteria will help prevent any unwanted surprises, such as traveller's diarrhea, which is usually caused not by infection but a change in microbial exposure.

This diet is an excellent foundation for other changes in lifestyle. For example, one of the fundamental requirements for optimal health is exercise, which should be performed every day. Those wishing to lose weight simply need to increase their level of exertion and the pounds will come off. If maintenance is the key, simply exercise less frequently. No change is needed in the diet itself to see results. It's that simple.

Herpesvirus

9. SEX

THE KISS OF DEATH It's absolutely incredible what happens when we share a romantic kiss. The lips tingle, the mouth fills with saliva, and a rush floods through the body. We may even begin to feel a rise in body temperature as more blood flows throughout. But all these sensations aren't always going to end up taking the relationship to a new level. A kiss, no matter how good in technique, may still be bad for a person, as it signifies the potential mate is actually not worth pursuing. When this happens, we may find ourselves choosing to shy away and hide.

The difference between hunger and distaste can be due in part to physical attraction. The combination of sight, smell, and touch are all contributing factors. But there is more going on inside the body as well. Two particular hormones, oxytocin and cortisol, play a major role in making a mating decision. Depending on which one is higher in concentration, you may decide to get it on or get on outta there.

Oxytocin is particularly involved in various forms of human relationships. It helps mothers during childbirth by promoting contractions, and later it encourages them to produce milk for their newborns. The chemical also helps new mothers develop empathy for and closeness to their children to increase bonding. In social interactions, oxytocin can amplify our feelings of closeness to another person, and during the mating process, it increases our excitement. (The ultimate contraction caused by oxytocin is the orgasm.)

Although this hormonal rush happens without much involvement from microbes, bacteria can play a role. Several friendly bacterial species in the gut will signal the body to stay balanced and may even send neurochemicals to the brain to amplify the concentration of the oxytocin. The result is an even stronger feeling of excitement and arousal. These bacteria can also help to reduce the inevitable low that comes once the oxytocin has stopped working.

Unfortunately, oxytocin is inferior to that other hormone, cortisol. It's released when we are stressed and can be a real buzzkill. When we feel under pressure, cortisol replaces romantic urges with a more determined but mundane effort to avoid harm, even if that harm is imagined. This can be made worse by those unfriendly bacteria in the gut, as they tend to ratchet up the hormone production. This in turn causes greater aversion and less interest. Introduce even more of these bad bugs and the cortisol levels could rise.

It may be hard to determine whether a person is ready for a kiss, although the scent coming from bacteria on the body and in the mouth may offer some clues. As I've said earlier, stress causes us to generate smells that can ward off even the most leering eyes. Also, stress can affect the bacterial population in the mouth, leaving us with the rather unpleasant odour of microbial by-products. The kiss could still happen, but it may not be enjoyed by either party.

For couples, this is a natural part of everyday life. But for those who are kissing for the first time, it could be the death knell for the relationship. Based on research conducted on over one thousand subjects in 2007, six out of ten people had at some point decided not to pursue a relationship simply because they'd had a bad kiss. One of the most common reasons given for disinterest was a lack of saliva and its associated symptom, bad breath, which itself is related to microbes. Granted, men were less choosy about breath quality, but women tended to look for more beneficial microbes—or if you wish, more pleasing breath—to continue kissing a potential mate.

FROM ME TO YOU It's quite incredible how much bacteria we share over the course of an intimate kiss. In as little as ten seconds, up to eighty million bacteria can end up in another person's mouth via saliva. That's the same amount of bacteria you would find in a standard serving of yogurt. A single serving of kisses won't make much of a difference to the overall microbial population of the mouth, but keeping up the kissing over a few days will allow bacteria to set up camp on the tongue. If you want to try to change the bacterial population of your partner's tongue, all you need to do is kiss that person at least nine times a day for a minimum of ten seconds. You may need to do it for only a few days to have an effect, but there's nothing stopping consenting couples from keeping this up over the long term.

WITH LOVE TO YOUR TEETH We can also share a number of infections by kissing. Colds, the flu, cold sores, gastrointestinal infections, and even mononucleosis can all find a new home in an unsuspecting partner.

But we don't seem to share problems with cavities and gum disease. When bacteria enter our mouths, they have a hard time getting to the teeth and gums. Even if they do, they will encounter an armed force of antimicrobial peptides, antibodies, and killing enzymes. These molecules are produced in the salivary glands and released whenever our lips touch another surface, whether it be the rim of a glass or another person's lips. Then there are the bacteria already in place, hogging up all the vacancy and unwilling to give away any room. They form a fantastic barrier that not even years of kissing can disrupt.

This all changes, mind you, if the mouth's defences are weakened in some way, perhaps by surgery or injury, giving tooth- and gum-haters a chance to attack. But even this can be avoided with proper oral hygiene.

A REALLY GOOD KISSER One virus that's passed on by kissing can have an entirely beneficial outcome. By infecting a woman with it, a lover may prevent a future child from suffering a potentially fatal infection.

The pathogen is called cytomegalovirus (CMV) because of its ability to make infected cells swell to a massive size.

It's present in over half of all adults, and causes fatigue, fever, and swollen glands. There are usually no further consequences. But if the virus gets into a fetus, there may be longlasting problems. CMV can interfere with development, causing deformations and even death. Infection of the unborn child has to be avoided at all costs.

CMV likes to hang out in the salivary glands, which is how it gets shared during kissing. If this happens while a woman is pregnant, the consequences can be dire. But if she is infected before conception, she may be able to develop immunity to the pathogen. This is where kissing may play an important anthropological role. A partner can actively share the virus during those passionate moments. The future mother becomes infected, but her body will develop the defences needed to keep the virus from causing any future harm. It's a form of vaccination, and there won't be any troubles once pregnancy starts. It's a simple way to disarm this particular virus should the couple choose to have a child in the future.

There are two catches to this method of child protection, however. First, only a small percentage of healthy individuals have the virus in their saliva. Second, it takes about six months for the future mother to develop immunity. The search is on for a vaccine against this virus, but it is still a ways off. In the meantime, you can be tested for the virus prior to conception to ensure the risks are low.

TINGLE BEFORE YOU TANGO As far as I'm concerned, it's always a good idea to have a shot or two of alcohol before sex. I'm referring of course to the kind you put on your hands.

Hand hygiene is always important, and especially so when we are about to touch someone's private parts. But taking that minute to wash your hands with soap and water right before lovemaking could seriously hamper the mood.

That's where a hand sanitizer can come into play. You don't need a sink or even running water—all you need is a quick squeeze and a good fifteen-second rub. In that time, the microbial species known to cause urinary tract infections and vaginal problems will be killed. The hands will be safe for intimacy. If you both use the sanitizer, then the certainty doubles.

You can even sneak the sanitizer into foreplay—it's a great excuse to rub those hands together and warm up the skin. Or if your partner has excused herself to slip into something more comfortable, there's more than enough time to clean your hands. You could even make it a part of the fun. There are pleasingly scented hand sanitizers to increase the overall atmosphere, and alcohol on the skin feels tingly and cool, which can increase the sensory excitement. Just don't put the sanitizer on anything too sensitive!

When you head out and get that special sanitizer for you and your loved one, make sure that what's inside the bottle is 62 to 70 percent ethanol. It's the only concentration that will be guaranteed to work. Apart from that, go out and enjoy!

ANOTHER GOOD REASON FOR PROTECTION There are many microbial benefits to using a condom—beyond the obvious and potentially life-saving one of preventing the spread of sexually-transmitted disease.

Condoms stop bacteria and viruses from going where they don't belong. For both men and women, this means the urinary tract, which is relatively microbe-free and likes to stay that way. As soon as microbes make their way inside the urinary tract, the defences immediately go about protecting their territory. Unprotected sex is the most common cause of urinary tract infections.

For women, there is another issue: the vagina is home to a very happy microbial population. When other microbes enter the area, particularly foes, a battle ensues. Normally, the residents can easily get rid of the invaders, but if weaknesses in the armour are present, the bacteria can quickly find a home and start establishing an infection. It's called bacterial vaginosis, is rather painful, and can last for weeks or months. Without proper treatment and complete removal of the bacteria, the condition can recur and remain persistent, even leading to troubles with fertility and carrying a child to term.

Men suffer infertility this way too, although much more rarely. Some bacteria can enter the urinary tract and make their way to the testes, where they attack and kill sperm. There are special species known to cause infertility, mostly without any symptoms, so that the problem is detected only when a fertility test is carried out. The source of the bacteria is feces. Without regular washing, bacteria can easily migrate from this nearby location and contaminate the genitals.

Condoms are imperative for those having casual sex, but the situation is different for monogamous couples. Once two people have decided to stay with each other—and are certain they're free of infections—forgoing the barrier might actually be a good thing. Couples in long-lasting relationships share microbes and develop a likeness for each other's friendly species. This can actually improve the overall enjoyment of sex because the shared bacteria will eventually be recognized as friends. The defences will not be needed, and pleasure can be enjoyed for years to come.

MICROBIAL MONOLOGUE The vagina is home to hundreds of millions of individual bacteria but a relatively small number of species. The most common are those lactic-acid bacteria. They find the warm, moist area the perfect place to thrive. They return the hospitality by ensuring that the acidity of the area is high enough to kill off any potential invaders.

These bacteria are not perfect, however, and cannot hold the fort forever. When pathogens such as yeast are introduced into the area, they immediately set to work lowering the acidity so they can stick around. And once those pathogens get a hold, they cause bacterial vaginosis. The cells of the vaginal walls are attacked, leading to abrasion, itchiness, and inflammation. The bacteria also produce odorous ammonia. This condition can recur, as some species are antibiotic-resistant.

Prevention of vaginosis isn't difficult. We know that the species causing the trouble are not usually associated with the human microbial population. That means they find their way into the area from the outside, by means of intercourse and unsanitary douching.

THE GERMS IN SPERM Most of the fluids in our bodies—from blood to urine—contain bacteria. Even semen has a microbial population, and it may determine whether a man is fertile or sterile.

Unlike the gut, semen holds very few distinct bacterial species, only a few dozen. This makes sense, as the only way the bacteria can get into the semen is through the little hole at the end of the urethra. Some bacteria are similar to the types found on the skin, and others normally find a home in the vagina. Most of these are harmless, but a few are potential foes. Some are sexually transmitted diseases (STDs), others are fecal bacteria, and a few others could cause vaginal infections in female partners. But the greatest threat comes from those capable of causing infertility.

Normally, these bacterial species have no bad intentions, but they do have a mean streak. They can cause both urinary tract infections and genital ulcers. For a man, this can spell trouble, as the bacteria can colonize in the testicles and ducts. As they continue their invasion, the body mounts its defence and takes out good sperm cells as collateral damage. Over time, the quality of the sperm decreases, and eventually, the man can become sterile.

Only a few species can cause this condition, but they are responsible for about 15 percent of all male infertility problems. As to prevention, good hygiene, particularly after bowel movements, will help keep the bacteria from getting into the urethra.

DUBIOUS DOUGLAS A few years ago, the actor Michael Douglas told the world he had throat cancer. And according to him, his condition wasn't caused by smoking or drinking.

"Without wanting to get too specific," he said, "this particular cancer is caused by HPV [human papillomavirus], which actually comes about from cunnilingus."

His claim seemed plausible. After all, a number of bacteria and viruses can be transferred from the genitals to the mouth and vice versa. Both areas have completely different microbial populations, and shifting their location can lead to infection. Usually, it's women having problems with vaginal infections, herpes and HPV. But men can also fall victim to viruses, including HIV, hepatitis, herpes, and HPV.

In reality, though, it's quite rare for HPV to be transmitted from the vagina to the mouth. At most, it occurs 5 percent of the time, and only in cases where an HPV-positive partner is in the process of shedding the virus. So while it's not impossible, it's quite unlikely.

But even if Douglas had caught HPV through cunnilingus, there's the issue regarding the development of cancer. The tonsils are the most likely place for cancer to start. The rest of the throat and even the mouth are at some risk, but it's quite low in the case of HPV infection. So either Douglas had an incredibly unlucky experience with microbes, or he was spinning a yarn.

It turns out the latter was true. Eventually, he admitted his cancer was not caused by HPV and indeed wasn't even located in the throat. It was on his tongue, and it was the result of a combination of alcohol and smoking, two of the leading causes of oral cancers. But even though his original story wasn't true, he did manage to bring attention to an underappreciated cause of sexually transmitted disease.

THE STAYING SIGNATURE OF SEX During the height of the AIDS pandemic, a certain phrase went viral: "You're sleeping with everyone your partner has ever slept with." The goal was to raise awareness of all sexually transmitted infections, and to promote the use of condoms. It was effective—the use of condoms increased worldwide.

But a condom may not prevent all spread of microbial life. The sheath covers only one part of the body, leaving the pubic area exposed. Because the hairs here are coarse and covered in sticky oils, they are the perfect place for bacteria to reside and thrive. When two bodies decide to grind together, the hairs intermingle. The physical movement alone is enough to force some bacteria from their current home to a new one. Add to that the natural lubrication of human sexual desire and you have the perfect medium for transfer.

Most of the time, the effort is wasted and the bacteria are removed during washing. But some of these species are more resilient and manage to stick around despite the introduction of soap and water. Once everything dries, the new inhabitants begin to make themselves at home. They can then be shared again when the person's desire turns to action.

This type of transfer isn't necessarily a bad thing. The majority of microbes shared are for the most part harmless. But they do leave a microbial signature. This can be a valuable forensic tool, particularly in cases of sexual assault. Prosecutors trying to prove that there was sexual contact need only take a clipping of the alleged offender's pubic hair and that of the victim. Within a day, the microbial populations of both can be compared. If there is a match, it's more than likely there was sexual contact.

This may also help to identify infidelity, particularly in heterosexual relationships. Men and women have different bacterial populations in their pubic hair, but long-lasting couples tend to share these microbes so their populations become more alike. Any new additions to the bacterial family could be a sign that a partner has been sharing microbes with a third person.

Still, the lab is no match for a private detective when it comes to collecting evidence for divorce court.

Staphylococcus epidermidis

10. CHILDCARE

THE VIRUS THAT MAKES US HUMAN Not long after conception, the fertilized egg implants into the uterine wall and begins to multiply. Some of the cells will become the fetus; others will form the umbilical cord. A third group is responsible for making a protective barrier around the developing child (the amniotic sac). The final group forms the placenta.

The placenta is a fascinating organ. It is completely separate from both the baby and the mother. It acts as a nutrient source for the fetus, removes waste, and forms a buffer against any chemical or microbial threats. Without it, no baby could survive.

Considering its importance in human propogation, you might think there is a specially designed mechanism in place to ensure a perfect placental formation. You would be correct, but there is a microbial twist. While mammalian cells make the placenta, the actual genetic code driving these cells is viral, namely a family of viruses called human endogenous retroviruses (HERVs).

HERVs don't exist in the same way as other viruses, such as influenza or Ebola. They don't actually multiply, spread, or cause disease. Instead, they sit in our cells in the form of DNA. When active, these viruses coax cells to produce a number of different proteins. Sometimes, these proteins can be harmful and may lead to psychological disorders and diseases such as multiple sclerosis. But in a few rare cases, such as placental formation, they are beneficial and needed.

The key to the viral contribution to the placenta is a protein called a syncytin. It forces cells to fuse together so they create a biological wall. As the fetus grows, syncytins are produced in greater numbers to ensure that the wall is kept firm. This process continues until birth, when the placenta is discarded from the uterus and eventually dies outside the body.

The importance of HERVs is not limited to humans; all mammals require the virally-encoded placenta. The introduction of HERVs may have been the critical moment in evolutionary history that allowed mammals to thrive. It might be a stretch to say that humans and other mammals would not exist without this virus, but it does seem to have become a necessity. This is just one of the many reminders that the relationship between ourselves and our viruses is a two-way street. We need them just as much as they need us.

BABY'S FIRST FRIENDS Pregnancy is an incredible phenomenon made even more amazing by the realization that everything happens in the confines of a relatively small space within the uterus called the amniotic cavity. It's made up of only a few parts, all of which serve a valuable purpose.

The amniotic sac surrounds the fetus to protect it from outside stressors. Inside, the amniotic fluid keeps the infant hydrated and also maintains a comfortable temperature. Just beside the sac is the placenta, which provides nourishment through the umbilical cord. Then there is the immune system, which protects the entire area with a collection of cells and chemicals to prevent unwanted foreign invasions, particularly from pathogens. After all, the presence of an infectious microbe can lead to dire consequences such as premature birth or worse.

The power of immune protection is so great that people once considered the amniotic cavity to be sterile. The only way a bacterium could enter the area, they believed, was if the immune system was compromised or the microbe had the ability to evade detection. But that's not the case. Throughout the entire pregnancy, bacteria are paying visits to the placenta and amniotic cavity, and sometimes they manage to get inside the child. We've since learned that this may actually be a good thing. Many of these bacteria help the child prepare for life in the outside world.

Why these microbes are allowed to visit this sacred place isn't exactly known. Are they just wayward hitchhikers, or do they serve an unknown purpose? They may be getting to know the child so they can be tolerated after birth. It makes sense. After all, they'll be there in the first few weeks to assist the baby in processing nutrients. They'll also serve as a barrier against hostile invaders. Some of these friendly bacteria also welcome other friends, increasing the diversity, which is so necessary later in life. It's as if their pre-birth visit is giving the child a head start on having the healthiest life possible.

HITCHING A RIDE Thousands of types of bacteria are constantly travelling around the body looking for new places to visit, such as the developing fetus. Some of them do this on their own—they are known as free floaters. But most have to hitchhike from the gut, the lungs, the sinuses, or the skin on what are called immature dendritic cells. The word *dendritic* is Greek for "tree," which is what these cells look like under a microscope. Normally a dendritic cell swallows a microbe and then breaks it down, killing it. But these immature versions don't have this ability, and so microscopic travellers can stay alive while they ride. When they reach their destination, the cell releases the microbe to continue its own journey.

Giving bacteria a lift helps the body get used to having these microbes around. Usually, their destination is a hub for the immune system, such as the lymph nodes, the thymus, or the spleen. When they arrive, these cellular taxis drop off their passengers, allowing them to come into contact with other immune cells. The visit doesn't usually last very long— most of these are mature dendritic cells, and they end up killing the bacteria shortly afterwards. This doesn't make for a very worthwhile journey for the bug, but for us, it's a fantastic way to get the whole body to recognize the various members of our microbial population.

NATURAL VS. CAESAREAN During natural childbirth, the baby travels through the uterus into the vagina. This area is known to have billions of microbes at any given time, but over the past nine months it has been getting ready by gathering and nurturing some two dozen special types of good germs.

These microbes are excellent at digesting nutrients, keeping the body hydrated and moist, and even protecting against other pathogenic bacteria. More importantly, they are recognized as friends. When the baby passes through the vaginal canal, where many of these same microbes reside, the very new and weak immune system welcomes them, making the first days of life as comfortable as possible. This introduction also helps to develop a base microbial population that will attract other friendly bacteria. A balanced diversity forms, comprising mainly symbiotic rather than pathogenic bacteria.

When a baby is delivered by C-section, the friendly microbes are not encountered. Instead, the baby is exposed to hundreds of germs it doesn't recognize. These include skin microbes from the delivery specialists, environmental microbes from the air and surrounding hospital equipment, and even some pathogens, which may cause infection and require antibiotics. The exposure to these unknown and potentially harmful entities puts pressure on the baby's almost non-existent immune system. It has no option but to consider them as foes.

This is a problem. In adults, an attack on a foe is a well-regulated and complex process with specific patterns to ensure victory. For a baby, the response is uncoordinated and cannot handle such invasions. Medication may help, but this does disservice to the immune system. It will simply not mature properly. The lack of a well-functioning immunity may cause an imbalance in the recognition and acceptance of friends.

In childhood, when most decisions are made regarding new microbial entries, fewer species may be given the right tolerance needed. This inevitably leads to imbalance and the risk for even more problems later in life, including diabetes, obesity, and arthritis during childhood.

Many doctors are now trying to figure out how to ensure that a baby delivered by C-section receives all the good germs from the vagina. One possibility involves the use of a swath of absorbent material. During a C-section, the pad is put into the vagina and all the wonderful liquids and microbes are collected. At the moment of birth, the infant is slathered with the vaginal fluid. The good bacteria are immediately transferred and are engaged to start protection. As they find their way to their new homes, they signal the immune system of their presence and help to hold off any inflammation.

THE MICROBIAL APRON STRINGS For the first few months of life, a child really shouldn't be around anyone but the mother and other family members. Their microbes are known to the baby's immune system and will not cause unneeded combat. But eventually, the child will begin to interact with new people, pets, and the wider environment (such as the backyard). In the process, an entirely new population of bacteria, viruses, and fungi will be discovered. Most will be bystanders, and some will be friendly. But inevitably there will also be foes, many of which can cause infectious diseases leading to illness.

The variety of new microbial visitors can present a challenge, as the population of microbes is far more diverse than at home. But it's well worth it. Children need to develop their own ability to recognize microbes and figure out the differences between the species. This knowledge is then stored away for life in the immune system's memory. This phenomenon, however, has nothing to do with the brain.

When the immune system comes into contact with any entity, the information is examined and the entity is designated as bystander, friend, foe, or self (meaning the object is actually a part of the body proper). That information is then stored in a certain population of immune cells, the regulatory cells I discussed in Chapter 1. They ensure that the defences do not carry out any unnecessary attacks. These cells also prevent the immune system from attacking the rest of the body, a condition known as autoimmunity. These overseers of the defence troops also ensure that any bacteria known to be friendly or at least harmless are left alone.

When a baby is born, the number of these regulatory cells is quite high, as they are seeking out and learning the environment outside the uterus. In the process, the bacteria are quickly assessed for their status. Those from the mother and direct family members are usually given a pass, as are several environmental microbes found in the home. If there are pets, they too can gain acceptance. As the process continues, the number of welcomed bacteria increases, leading to greater diversity.

But there is a drawback to this process. While the regulatory cells are gaining information, they keep the immune forces suppressed. This ensures that the troops don't end up hurting the growing baby's own cells as she grows. But this reduction in response also means that foes have an opportunity to cause illness. Although the defences may be able to take out the invader, the regulatory cells keep them from doing their worst. The child can end up sick and in need of medical attention.

The process of regulation in the newborn takes about six months, after which the baby can safely encounter people outside the immediate family, animals, and other microbial sources, like mud. But prior to this, exposures should be limited to friendly and harmless species. Unfortunately, that's easier said than done. Unless a person is overtly ill—with a cold or the flu, say—it's hard to judge if the microbes being passed on will be friendly or harmful. The same goes for animals. They can harbour bacteria in their saliva, dander, and fecal matter. Many are harmless, but some can cause illness and spread to an unsuspecting baby.

To keep the child safe, visits needs to be carefully controlled. Members of the extended family should exercise caution, and definitely wash their hands before touching the child. For those who come from outside the family, there really is only one rule: look but don't touch. Many visitors will want to dive in for a whiff of new baby or hold the child for instant gratification. But for the sake of the child's health, a parent should wait until that immune system is better developed before handing baby over to a microbial stranger.

Proper hygiene is also important for pets. Though they may help to increase the child's microbial diversity, they may also cause some harm in the first few months. Washing the dog regularly and disposing of kitty litter can ensure the risks are kept low. Any inadvertent bites and scratches should be cleaned and examined for signs of redness and soreness; contacting a medical professional is also good practice. Finally, make sure all animals are up to date on their own visits to the vet—this will keep pets *and* baby healthy and safe.

VACCINES DO NOT CAUSE AUTISM Vaccination has been controversial ever since the first one was given over two hundred years ago. Doctors and patients have expressed concerns about side effects. This resistance has for the most part been beneficial, as it has led to the development of ever-safer vaccines. There is, however, one allegation against vaccination that is entirely without merit: no matter what people think, vaccines do not cause autism or autism spectrum disorders.

Unfortunately, this particular allegation—unlike most unfounded complaints against medicine—was started and propagated by a scientist. His name was Andrew Wakefield, and in 1998, he found a hint that children were at a higher risk for developmental disorders if they received the measles, mumps, and rubella vaccine.

It was, however, just a hint. All Wakefield did was count heads and try to match autism with medical histories. His evidence was circumstantial at best, and he admitted this, explicitly saying in the paper that there was no conclusive proof to support the claim. But that didn't stop him from calling out the vaccine for its supposed links to autism and gaining an army of support in those who feared immunization.

After Wakefield's paper was published, researchers attempted to confirm the results. But they found no smoking gun. Nothing could lend the support needed to prove that vaccines caused autism. Head counts were not the solution; to get to the real answer, researchers had to go under the microscope. The efforts revealed the complexity of autism as a whole and the reasons why vaccines play no part in the disorder.

There is no single trigger for autism. It's a combination of genetic and environmental factors, such as diet, exposure to toxic chemicals, medical therapies such as antibiotics, and the nature of the various bacteria in the gut. Most of these factors have no link to one another. But there is one common denominator: each of these triggers has a link to the immune system and can cause an imbalance leading to inflammation. When this happens in the brain, cells may be harmed. In a growing child, this can stall or even stop proper development and increase the chance for alterations in physiological and psychological function.

Vaccines are designed to work with the immune system. They introduce a harmless version of a pathogen to set off a controlled response. Granted, inflammation inevitably occurs at site of the injection, causing some pain and soreness. There may also be some systemic reaction, such as a fever. But the brain is protected from any damage. The only cerebral consequence may be the occasional visit from the newly formed antibodies. But these immune soldiers do not cause harm or contribute to delayed development.

WHY BREAST IS BEST As everyone knows, the perfect food for a newborn is mother's milk. As I said earlier, it has a high level of antimicrobial agents capable of keeping the gut safe from pathogens. But there's so much more that makes breast best.

Much like the milk of other animals, human milk contains lactose, which is easy to digest and offers much-needed energy. There are also longer types of sugars, known as oligosaccharides.

These take longer to break down and so sustain the energy in between feedings. Fats and proteins within the white liquid act as building blocks for new cells and increases in organ size. Along with them, there are certain hormones, called growth factors, which stimulate the development of the body from blood vessels to organs to the nervous system. Then there are vitamins to keep the body running smoothly. But this is where the similarities end and mother's milk asserts its superiority.

Breast milk contains molecules necessary to defend the baby against many microbial foes. There are those lysozyme enzymes we looked at earlier, but there are also a number of other needed molecules to help baby stay healthy. These include hormones, to keep the integrity of the gastrointestinal tract; antibodies, which seek out harmful microbes and block them from causing any damage; antimicrobial peptides, to kill pathogens and neutralize their toxins; and cytokines, which are chemicals known to signal the immune system and keep the fragile defence forces calm but alert to potential danger.

There is a final benefit: bacteria. Breast milk is not sterile, but instead is populated with species that are friends with the mother. These take up residence in the child, adding to a growing diversity that started at birth.

There are only a few dozen types of microbes in human milk, and they come from all parts of the body, including the mouth, the skin, and the gut. How they get to the breast is still being debated, but it is believed they hitchhike to the mammary glands on the backs of human cells. However it happens, these bacteria offer a child a safe circle of friends.

The benefits of breast milk are probably best understood by looking at babies who are not breastfed. The types of bacteria found in their gut can vary significantly, which sometimes leads to more infections. But this isn't the only problem. As the child grows, immune system imbalances caused by a lack of microbial friends can allow for the development of allergies and asthma. The lack of these good microbes may also hinder the metabolism of nutrients. This could lead to changes in the way children eat, leaving them prone to obesity and diabetes.

Not everyone chooses to breastfeed or is able to do it so some manufacturers of baby formula are adding probiotics. In the coming years, more species may end up in the mix. Clinical trials suggest that certain non-probiotic species can ward off ailments such as asthma and allergies later in life. Rigorous testing needs to be carried out before these species are designated as probiotics, but one day they may enable children to grow up without any respiratory troubles.

WHEN TO WEAN? It's a question I have been asked quite often by mothers: "How long should I breastfeed my baby?" What they really want to know is how long does a child need to be breastfed to develop a diverse microbial population that will help maintain health in later life. The answer is relatively simple: at least six months, but preferably a year. The bacterial population of the gut takes about six months to develop and then another six months to stabilize. During that time, the breast milk will keep the population balanced with a good proportion of friendly bacteria. The milk will also provide many immune factors to help prevent species of harmful bacteria from causing trouble. By the one-year mark, the immune system will be able to take care of the child on its own.

BABY TAR For any new and unsuspecting parent, baby's first bowel movements can be quite a surprise. They are black, tarry, sticky, and for the most part have no smell. Some new parents take this as a sign that something is wrong. But the unsightly substance is quite normal and plays a role in the transition from the womb to the world.

The mixture is known as meconium (from the Greek word for poppy juice), and it is a collection of waste products formed during pregnancy. When the baby is comfortably floating in amniotic fluid, there is no proper route to get rid of dead cells, mucus, amniotic fluid, and hair. Instead, these things are collected in the gastrointestinal tract and sit until they can be released. Over the first few days of life, the meconium slowly works its way out of the body to allow for other nutrients, such as breast milk, to take its place.

At one time it was thought that babies were born free of microbes, and that meconium was sterile. But this isn't the case. Microbes thrive on meconium, which contains all the nourishment they need to grow. By the time the baby is born, the bacterial population can reach the trillions. Although we don't entirely understand why those bacteria are in the fetus in the first place, we do know that they are an indicator of the potential for a healthy childhood.

In healthy babies, the meconium contains a diverse population of bacteria from various areas of the mother's body.

These microbes are transported to the uterus and from there through the placenta or amniotic sac to the fetus. Most of these bacteria do not pose any threat. All they want to do is feed on simple sugars and return the favour by producing beneficial chemicals and molecules that help to keep the immune environment calm. This latter benefit is quite important for the fetus, as its immune system is not well developed and can use all the friends it can get.

When the baby is born, some microbes decide to hitch a ride out of the body, but many stick around and become part of the baby's gastrointestinal bacterial population. This is possibly because they love the sugars in breast milk. After the first feeding, they decide to take up a more permanent residence.

That all changes if the number of good bacteria falls and the number of foes rises, which could happen if the mother is given antibiotics or smokes or has a poor diet. If unfriendly microbes gain ground in the meconium, the entire uterine environment goes into a state of immune activation, which could lead to premature labour. That premature baby is then at risk from infections and a potentially deadly condition known as necrotizing entercolitis. Both are treatable with antibiotics, but for a newborn this can be a severe shock to the system, leading to continued imbalance for years to come.

Unfortunately, it is not possible to measure the meconium while the baby is still in the womb. Possible problems can be identified only through analysis of the amniotic fluid. However, taking a look at the microbes in the tarry output of a newborn child may offer insights into her overall health. It may also help to identify what microbes are missing and need to be supplied to prevent potentially horrific consequences during those first few weeks of life.

50 SHADES OF DIAPER You can tell quite a bit from the colour of a baby's feces. Taking a few moments during a diaper change to note the hue and consistency can provide reassurance about the child's health or act as an early warning.

The usual brown colour of feces comes from the mixture of two bodily fluids: bile, which is yellow and helps in digestion, and bilirubin, a yellow-orange-brown waste product formed when our red blood cells die off naturally. (Bilirubin is also the chemical responsible for turning bruises that sickly yellowish tinge.) But the colour of feces can change depending on our diets. Beets will give a red tinge thanks to the presence of chemicals called betalains. A diet high in green vegetables will tint the stool dark green or even black, in part because of their high iron content. If you eat lots of carrots or take beta-carotene supplements, an orange hue may appear. Then there are those food dyes, which tend not to break down in digestion. If you've recently had grape soda or a cocktail made with Blue Curaçao, you might notice your stool is blue.

For a baby, fecal colour provides a valuable indication of the state of the microbes inside the gut. Depending on what the baby is being fed—breast milk or formula—a number of different hues are possible. The colour provides an indication not only of the digestive process but also of whether the bacteria are doing a good job.

Solely breastfed infants very often pass bright green stools a few days after birth. This is a result of bile and is a sign the newborn is getting ready to digest foods. Within a few days, the colour shifts dramatically to pale yellow. This hue is from the multitudes of bacteria sent out into the feces. The fecal matter may also appear to have lumps in it, but don't worry: this is due to milk fermentation, much like we see in sour milk. As the child grows, bilirubin will be produced, darkening the pale yellow to a more mustard hue.

Once formula is introduced, the browns really begin to kick in. At first, the feces will resemble peanut butter, guacamole, or hummus as the range of hues can incorporate tans and greens as well. This shows the bacteria inside are adapting well to the new food source and digesting properly. The smell should also become less sweet and more like, well, poop. This is owing to the formation of microbial by-products important in maintaining well-balanced metabolic and immune systems. If iron supplements are taken at this time, the stool will darken even further and may even appear black.

If the stool has any other colour, it could indicate a problem. Red is the colour to take most seriously. This is an indication of blood in the stool and a possible injury, and needs to be looked at by a doctor as soon a possible. If the injury is on the inside, bacteria may have entered the bloodstream, which can lead to a serious condition known as sepsis. Black stools— although common in the first few days of life—should also be treated as a sign of a potential problem. The cause is a high iron content. If the baby is not on supplements, then the iron is coming from blood high up in the gastrointestinal tract.

This means there is a tear somewhere deep inside the baby's intestines, and it needs to be checked out right away.

White feces are an indication of another major problem: a lack of bile in the gut, and therefore poor digestion. This also needs to be checked right away. It could signify a problem with the liver. The bacteria inside the gut also could be affected, as the bile maintains a healthy environment for good bacteria. Bile not only helps digest food but also kills pathogens. Without this vital fluid, certain bacteria known to cause diarrhea and pain may end up growing freely inside the child. This will compound the problems already caused in the liver and require even more intensive medical attention. Essentially, if the stool is white, call in the white coats.

By the time the child reaches six months and is beginning to eat solid food, the colour wheel may no longer apply. The baby's gut is already becoming more adult. The feces turn brown and the smell becomes stronger. Thankfully, those extra few moments spent checking out the stool colour may no longer be necessary.

TIME FOR A CHANGE? Disposable, absorbent diapers are a boon for modern parents, but they may not be quite such a good idea for babies, particularly baby girls. Liquid gets absorbed, but enough is left behind for bacteria to grow. If any fecal bacteria happen to be around, they can take up residence and spread to the urinary tract. The end result is an infection that is both painful and also may be difficult to treat, particularly if the bacteria are antibiotic-resistant.

The way to control this risk is to limit the use of absorbent diapers, particularly those described as super- or ultra-absorbent. At first glance, this may seem illogical, as super-absorbency should lead to less liquid. It does. But it doesn't get rid of it all, and that's the problem. The best means of drying the genital area is through ventilation, but many super-absorbent diapers simply do not allow much air to get in. This allows liquid to pool, giving those bacteria a chance to cause troubles.

Cotton diapers are the best, but most people won't use them. So compromise may be the best way to keep baby safe: use regular absorbent diapers with ventilation incorporated into the design to allow them to fully dry.

THE CAUSE OF COLIC Babies cry to tell us they are hungry, tired, in need of a diaper change, unwell, lonely, or even bored. And in one case they cry—and cry, and cry—to let us know there is something amiss with their resident microbes. The problem is known as colic.

The name comes from the Latin word *colon*, and the colon is where the troubles originate. In the first year of life, approximately a quarter of all babies will suffer at least once from this gastrointestinal condition. It causes cramping, diarrhea, and continuous pain. Hence the incessant crying, which no amount of outside comfort seems to help. Resolving the issue can take days, if not weeks.

We still don't fully understand what causes colic, but there is no doubt that children who have it have a very different microbial population from children who don't. First off, the balance of microbes in colicky children is lower on friends and higher on foes. These microbes are gluttons for nutrients and will fight to get them. Whey they do, they form a variety of by-products that harm the resident bacteria and the human cells, causing gas, a bloated belly, and even pain.

The standard treatment is medication to prevent gas. This is only a Band-Aid solution, but for a worried and sleep-deprived parent, any respite is welcome. Pain medications including anti-inflammatories may also help, but these too are short-term options. It's probably not a good idea to administer them for as long as the colic exists.

Antibiotics may also be used, but they can disrupt the bacterial population of the gut and lead to an imbalance. Though the foes will be taken out in the short term—that is, if they are not resistant—the intestinal landscape becomes barren. Once the prescription is done, the gut becomes prime territory for colonization by any and all bacteria, including several unfriendly species. If this is the route taken, it's critical to introduce good bacteria, either through breast milk or through supplements like specially designed probiotics. There are currently a few probiotic options for colic. But because of the fragile nature of an infant's health, these supplements should only be sought after consultation with a doctor.

A PAIN IN THE EARS Earache is a common problem for babies and young children, and an excruciating one. I should know, as I suffered from them constantly as a kid. The doctor's office was a familiar place for me, and at times I wondered why my ears hated me.

I had no idea back then that the source of trouble wasn't always the ear canal. It also involved other regions where bacteria care to roam, including the nose, sinuses, and throat. An earache, it turned out, could have numerous causes, from bacteria to viruses to fungi.

A few of these happen to be normal microbial members of the respiratory tract. Most of the time, these species are friends. But that changes when they suddenly turn face and end up adding to the trouble. For the defences this defection is a shock, and they need time to figure out how to respond. As this happens, the ear becomes a feeding ground for the bacteria to grow and establish a colony. Eventually, the immune system figures out what to do and engages. But by this time, the battle will not be quick and simple. It can take days or even longer and sometimes requires medicinal help in the form of antibiotics.

The situation is made worse when these bacteria seek out other areas, including the adenoid glands and the tonsils. Here, they can hide out for years, waiting for the right moment to attack. This can result in recurring infections and more regular trips to the doctor for antibiotics.

There is one course of action for children with chronic ear problems: the adenoids and tonsils can be removed. But even then, the risk for future problems may not be eliminated. There will still be high levels of these same bacteria in the nose and throat, suggesting an earache could happen at any time. The best way to avoid these sporadic flare-ups is to maintain a healthy ear canal and prevent the introduction of pathogens. This can include regular washing and careful drying of the outside ear (never the inside!) with a cotton swab.

Improving the mix of bacteria in the nose and throat will also help reduce the risk. One of the best ways to do this is to practise good oral hygiene; the mouth is a perfect entrance to the rest of the respiratory tract. The introduction of friendly bacteria in the form of drops or gums can also help. They can help prevent unfriendly bacteria from invading and keep other friends from turning into foes.

RED ALERT Like any other place on the body, the eye has a bacterial population. Most arrive from the skin or the air. They don't head straight onto the cornea or the liquid underneath, but instead find a home on the protective layer of mucus covering the eye.

In the eyes, mucus covers the cornea, the most sensitive part of the eye. Here, everything is kept lubricated and moist. But the mucus is not free of microbes. Several bacterial species find a happy home there. This isn't a bad thing, as these inhabitants serve a dual purpose in maintaining eye health. The first is to act as recyclers. During our day-to-day lives, our eyes create quite a bit of waste, from unused nutrients to that slimy mucus. Although we can remove them with tears, the bacteria offer help by eating up these unwanted molecules. The second is to offer increased protection against pathogens. These bacteria produce a number of chemicals, known as antimicrobial peptides, to kill our eyes' enemies. By taking care of any threats before our immune systems has to kick in, the bacteria can forestall potential problems and reduce our need to constantly tear up.

There is, however, a limit to the acceptance of these bacteria. If for some reason they begin to gorge on the mucus lining and make it to the cornea surface, all bets are off. Even though these bacteria are our friends, our immune systems are trained to identify any potential threat. It sets off a cascade to fend off the invader.

As soon as the cornea surface is touched by the microbe, cells and proteins rush to the area to engage the intruder. As they do, they release other molecules, such as the itchiness-causing histamine, to let the rest of the body know what is happening. A rush of fluid (tears) enters the area to dilute the enemy and weaken any possible strike. The environment also becomes inflamed. The immune cells attack and kill any microbes not removed by the tears. The entire process may take only a few minutes, but the redness from the inflammation can take hours to recede.

Normally, the process is as quick as it is vicious. But some of those bacteria are more willing to put up a fight. Instead of going away, these species will do their best to adhere to the cornea and invade even further. When this happens, the inflammation spreads and the entire area becomes red or pink. If not given medical attention, the inflammation may cause keratoconjunctivitis as well as other complications, including scarring and the formation of thickened regions of the cellular tissue, hindering overall sight.

The difference between an inadvertent microbial visitation and an attack can be difficult to discern, as both end up causing the same problems in the short term. But if the redness remains after a few hours, it might be a good idea to contact the doctor, if only to assess the situation. The one thing you don't want to do is give eyedrops to children. Though these are great for adult eyes, they may cause little ones unneeded pain and possibly disrupt the already delicate balance among cornea, mucus, and microbe.

LET THEM EAT DIRT Mud is a kid-magnet. That's usually a good thing. Unless contaminated in some way—say, by animal feces—the bacteria that call the wet soil home are either harmless or positively friendly. As soon as a kid jumps in, the mud will convey these species onto the skin, into the respiratory tract, and even into the gut. The child becomes one with nature and its incredible microbial diversity.

But this isn't the only reason mud is so good for kids. It may also help a child's mental health, providing long moments of joy and entertainment. This too has a microbial origin. A particular bacterial species found in the soil is known to make mammals feel happy. It's called *Mycobacterium vaccae*, meaning it is normally found in fungus and also in cows. It somehow sends our minds signals of happiness and joy. No one quite knows exactly how this happens, but the end result is always the same—mud makes for merriment.

There's a final benefit to getting all mucky, although it has little to do with the microbes in the soil. The overall elation we feel when we get dirty tends to also keep a balance with the bacteria in our gut.

For anyone dying to get dirty, there is more than enough evidence to support the value of having a good time in the mud. And parents have an excellent excuse because they are simply trying to keep their children safe. The rest of us can seek out more refined places such as spas with mud baths in order to enjoy a few moments of dirty bliss.

NEVER TOO YOUNG TO WASH HANDS I'm often asked by parents when a child should learn about handwashing. I always respond, "As soon as they are out of the womb." This comment is usually met with a curious look and the occasional "You're not a parent, are you?"

But when it comes to hand hygiene, you're really never too young to take part. I do realize that children may not actually memorize how to perform handwashing until much later on—anywhere between eighteen months and two years of age—but they will nevertheless appreciate the routine as a part of their instinctual selves long before that.

The best way to raise a handwash-loving child is to remember to wash your own hands as often as possible. This can be after changing a diaper, after feeding, and after returning home from outside. First, wash your own hands and then wash the little one's hands. You don't have to use a sink—even a small glass of water will do. You can use the same soap you use for bathing baby. After all, this is about developing a routine, not necessarily getting rid of any pathogens.

As children grow, they will become more aware and can start taking on some of these actions by themselves. This might mean putting their hands in the running water or maybe even trying to play with the soap. Each additional step will be a milestone, until finally the little ones will be able to do the entire routine by themselves. If all works well, this will happen at about the same time as (or even before) potty training, when the handwashing will be needed.

There is one form of hand hygiene that should be delayed, and that is the use of alcohol sanitizers. The concentration in these gels is quite high and the taste is horrid. You should wait to introduce these handwashing supplements until the child is old enough to appreciate the action of the rub without having to taste whatever happens to be poured. For most children, this is around the two-year mark. But there's nothing wrong with waiting even longer than that. After all, if a child already has handwashing down pat, it'll be some time before she reaches an age when a sanitizer is needed.

THE DAYCARE DILEMMA Nurseries and daycares are microcosms of the microbial world. Each child carries hundreds of different species, many of them sharable with the other children. Add to that the variety of ethnic and cultural backgrounds, and suddenly attending daycare amounts to a global diversity of microbes. This fact alone means your child is at risk for colds, the flu, gastrointestinal distress, pink eye, and earaches. The silver lining is that the child's immune system develops a deeper understanding of the microbial world.

For kids, getting sick is a training exercise for a branch of the immune defence force, the one tasked with memorising how to fight infections. When trained, this force can remember bacterial and viral threats, and even eliminate an infectious agent before any symptoms occur.

But of course the child has to become infected for the initial training to begin. And that means symptoms. The onset of fever is the immune system's typical response. Depending on how strong the defences need to be to control the infection, other symptoms can include diarrhea, nausea, vomiting, coughing, and sneezing. Sometimes medical attention may be sought, if only to ensure that the child is coping well.

In the process of responding to an infection, the immune system gains insight into how the pathogen infects cells, spreads in the body, and damages the cells and tissues. Somewhere within this process will be a weak link that can be exploited to stop the enemy. It could be a structural component, like a part of the outside wall or a necessary protein or even a valuable sugar. One way or another, the infection has a weakness and the immune system finds it. When it does, the infection ends shortly afterwards. Once the symptoms subside, the child should be protected against that pathogen for life.

There's a catch. Some pathogens have the ability to block or evade memory and can re-infect. These are mainly viruses such as the common cold, rhinovirus, influenza, and the gastrointestinal norovirus. Children who attend daycare centres are more likely to acquire one of these, but without long-term benefit. Keeping these infections out is mandatory and requires adherence to proper hygiene. Thorough handwashing and disinfection of surfaces will ensure minimal spread and less unnecessary illness.

TROUBLE AFOOT It's a gripe I have heard from parents for years. When it's cold outside, their children are adamant they don't need to wear warm shoes or boots. So the exasperated parents forewarn a microbial consequence: "You'll catch a cold." This is not an empty threat; it's actually quite true. Except that the common cold isn't actually caused by having cold feet. Instead, the cold takes advantage of a certain trait we humans have evolved over the millennia.

When our bodies get cold, our blood vessels constrict in order to keep our blood warm. Even though only one part of the body is chilled, the effect is spread throughout. As a result, the defences of the immune system are not allowed free access to all areas, particularly the sinuses and upper respiratory tract. In the event a virus happens to be present—or is introduced shortly after the chill—it is left alone and not attacked. Sensing the opportunity, the virus quickly launches a campaign to cause infection and leave us with the sniffles, sneezes, coughs, and other cold-like symptoms.

The risk for catching a cold-feet cold is, of course, dependent on whether viruses are actually circulating during that time period. But it's quite well known that these pathogens tend to be seasonal and are most prevalent during the cold months of the winter and early spring. It's why parents should do all they can to ensure their children's feet are warm and cozy during that time of the year, even if it's not, well, cool to do so.

GERMOPHOBES TO GERM PHILES There are aspects of human nature we hope not to pass on to our children. A lack of concern for the environment is one; we hope our children will be more respectful of our home. Discrimination is another, as we look to our children to make the world more tolerant and accepting. We also try to discourage certain habits, such as smoking, so the next generation won't end up on a path of self-destruction.

There's one more troublesome trend that should never be passed on to those who follow us: it's time to end germophobia. Fear of microbes may have been valid two or three generations ago, but we need to change this mindset now. Already our inability to accept and respect these microbes has had significant consequences. Our environment is plagued with unnecessary excesses of antibiotics. We've discriminated against all species for the sake of only a few bad actors, leaving us with chronic health issues such as dysbiosis. Then there are the bad habits we've gained—buildings without operable windows, overuse of cleaning chemicals, and eschewing nature for the urban jungle.

As we look at those who will be in charge of our world, we need to instil within them a love of germs. They have to understand that the majority of microbes are beneficial to us and the earth. Yes, pathogens deserve tough love in the form of banishment. But this action also has to be measured.

Raising a germphile isn't as hard as it sounds. After all, germs offer far more good than bad. Granted, in the first few years, parents need to be vigilant about anti-pathogen actions, such as careful hygiene and the occasional visit to the doctor. But with each lesson in prevention, parents should also offer one on the benefits of microbes. A balanced perspective can make a child aware of the troubles but also ready to commit to the good that comes from our microbial friends.

If your children are old enough—and you are okay with them reading chapter 9—you may even want to pass on this book for their interest and pleasure. I'm sure it will help in the development of a germ-loving mentality, and perhaps even set the stage for a career in microbiology so we can all look forward to even more fascinating microbial finds.

ACKNOWLEDGMENTS

I continue to be thankful to all media organizations that have adopted me as their "Germ Guy." Without them, I would simply be another voice in the crowd. I am especially grateful to Marie Clarke-Davies at the CBC, Haley Mick at the *Globe and Mail*, Sandie Rinaldo at the CTV Television Network, Jessie Lorraine at CFRB radio, and Terry Mercury at Sirius XM Canada. Their calls will never be taken for granted. Thanks also to Ward Anderson and Allison Dore, hosts of the *Ward and Al Show* on Canada Talks Radio, who allow me to adopt a more comedic style, complete with my Horatio Caine–inspired one-liners and updates on the status of the coming zombie apocalypse.

I owe significant thanks to Jenna Elfman. Though she is best known as a Hollywood actress, she is also an incredible, supportive sage who gave me a few useful virtual kicks in the posterior while I tackled this book.

I would again like to thank my editor at Doubleday Canada, Tim Rostron, for his guidance and unending patience, as well as Scott Sellers, associate publisher at Penguin Random House of Canada, for his continued enthusiastic support. Thanks also to freelance copyeditor Janice Weaver; Melanie Tutino, Doubleday Canada editorial assistant; and editorial intern Nandini Thaker for their eagle eyes and thoughtful comments.

Big thanks to my parents, Peter and Patricia Tetro, and my good friend Jason Gilbert. They always believed in me, even when I went through the toughest of times. Their encouragement and consistent demands for clarity provided me with the drive to carry on and make sure what I wrote was meaningful to those outside the lab.

Thank you to everyone who reads, listens to, and watches my work as the Germ Guy. Your support gives me so much energy each and every day.

And above all, I want to thank my love and the first reader of all my work, Anastassia Voronova. She has guided and inspired me to be a better scientist, a better communicator, and (it may sound a little clichéd, but it's so true) a better person.

INDEX

Jason Tetro has called the microbiology lab his second home for a quarter of a century. He studied at the University of Guelph, and for fifteen years researched health-related microbiology at the University of Ottawa. He has been a consultant to government and private industry in many health-related areas, including food safety, blood-borne pathogens, infection prevention and control, and immunology. He is a regular contributor to print and television media, and a prolific public speaker. He lives in Toronto.